TEMPERATURE

MEASUREMENT

IN

ENGINEERING

TEMPERATURE MEASUREMENT IN ENGINEERING

VOLUME I

H. DEAN BAKER, Ph.D.

Professor of Mechanical Engineering
Columbia University
Consultant, Pratt & Whitney Aircraft Division
United Aircraft Corp., East Hartford, Conn.

E. A. RYDER, M.E.

Consulting Engineer, Pratt & Whitney
Aircraft Division, United Aircraft Corp.
East Hartford, Conn.

N. H. BAKER, M.A.

Research Assistant in the Department
of Mechanical Engineering
Columbia University

OMEGA PRESS, a division of
OMEGA ENGINEERING, INC.
STAMFORD, CONN.

COPYRIGHT 1975 by OMEGA ENGINEERING, INC., STAMFORD, CT.

Copyright, 1953, by John Wiley & Sons, Inc.

All rights reserved. This book or any part thereof must not be reproduced in any form without the written permission of the publisher.

Some of the material in this book is from the author's Manual on Thermometry, copyright 1950 by The United Aircraft Corporation, East Hartford, Conn.

Library of Congress Catalog Card Number: 53–11565

Printed in the United States of America

PREFACE

The present book has grown out of a research and development project prescribed by the late Dr. Charles Edward Lucke as part of a general research program. The work was conducted in the laboratories of the Mechanical Engineering and Physics Departments of Columbia University. The success of this project was in no small part due to the wise guidance of Dr. Lucke, who had long recognized the vital importance of adequate temperature-measurement technique in engineering.

The object in this two-volume work is to provide in convenient nonmathematical form the information necessary to the engineer who wishes to measure temperature. An effort has been made to include all those specific details essential to actual execution, while refraining from encumbering the space with material of infrequent application or of the nature of background theory. It is recognized that each actual measurement tends to be a problem in itself. The approach here has been to provide a comprehensive list of possible techniques, the methods of analysis, a survey of previous designs, the specific information required for feasibility of execution, and a well-developed procedure of general applicability.

A compact book, intended for convenient reference, cannot contain all available material of practical importance in relation to each of the myriads of specific problems in temperature measurement. Much of this is too cumbersome and related to narrow specialties; some of it properly belongs in separate books; parts of it are continually changing. In order that this supplementary information for each specific case be conveniently available, complete reference data has been given. Superior numbers in the text indicate the points at which additional reading is recommended. Also, the listings at the ends of chapters include the titles of specific references, to assist the reader in estimating their pertinence to his immediate interest.

In contrast to previously published books on temperature measurement, stress is placed on the specific procedures and techniques involved in producing satisfactory temperature-measurement designs for various circumstances. The types of conditions encountered are classified on a physical basis: as interior points in solids, liquids, gases, flames, etc., rather than as the problems of specific industries or as related to various types of instrumentation.

Volume I deals primarily with thermocouple technique, only because this is the most widely useful method of measuring internal temperatures of solid bodies. The first four chapters are introductory, with respect to the two-volume sequence.

Volume II deals with problems in the measurement of very low and very high temperatures: surface temperatures; temperatures in rapidly moving bodies; temperatures of liquids, stationary and high-velocity gases, and gases carrying entrained particles of liquids and solids; and temperatures of flames.

For the situations covered in Volume II, thermoelectric technique does not have the unchallenged superiority that it offers for internal temperatures in solids, as discussed in Volume I. Individual problems necessitate specific techniques. In consequence, much space is devoted to a variety of other devices for temperature measurement as these are necessitated by the circumstances encountered. Thus, electric-resistance thermometry, the velocity-of-sound and orifice methods of temperature measurement, high-velocity aspirated thermocouples, many special forms of radiation pyrometry, and various other devices are treated.

Whereas many of the chapters are essentially original and consist of material not printed elsewhere, free use has been made of manufacturers' literature and of previously published matter. Since references are intended to direct the reader to additional useful information rather than to indicate the source of that already given in the book, indebtedness to previous work is not uniformly acknowledged. Particular assistance is recognized, as from *The Modern Calorimeter* by W. P. White and *Liquid-in-Glass Thermometers* by J. Busse.

Although the persons associated with this book are too numerous to list in full, particular credit is due to W. Claypoole, who performed much of the actual test work, E. D. Brown, F. Freudenstein, W. J. Peets, S. A. Ochs, G. L. Laserson, M. Mage, and M. Petschelt, each of whom made important contributions, and to R. H. Wilson, who typed the manuscript.

H. Dean Baker
E. A. Ryder
N. H. Baker

New York
September, 1953

CONTENTS

1 · Temperature	1
2 · Methods for Measuring Temperature	13
3 · Precision Requirements	23
4 · Conditions Affecting Temperature Measurement	30
5 · The Thermocouple Thermometer—Circuits	34
6 · Indicating Instruments	64
7 · Design Calculation Techniques	68
8 · Installation Design Types	84
9 · Drilling Technique	107
10 · Special Materials: Protective Coatings, Heat- and Corrosion-Resistant Metals, Plastics, Refractories, and Cements	128
11 · Cemented Installation Designs	142
12 · Temperature Gradient Installation Designs	159
13 · Conclusion to Volume I	168
Name Index	169
Subject Index	173

1

TEMPERATURE

1·1 THE TEMPERATURE CONCEPT

The concept of "hotness" or "coldness" originates in the sense of touch. This human sensation can be correlated with various objectively discernible changes in physical bodies, such as expansion on heating and contraction on cooling, changes in resistance to the flow of electric currents, and luminosity of incandescent bodies. These effects can be used as means of measuring relative degrees of "hotness" or "coldness"—in other words, to measure *temperature* and to establish *temperature scales*.

1·2 THERMOMETERS

Various types of *thermometers*, or temperature-measuring instruments, are based on such physical changes dependent on temperature. The more familiar among these are the liquid-in-glass thermometer, the thermocouple, the electrical-resistance thermometer, the optical pyrometer, the Bourdon gas or vapor-pressure thermometer, pyrometric cones, and bimetallic strips.

1·3 THE MERCURY-IN-GLASS SCALE

Pure materials, such as distilled water, melt and boil at definite temperatures, provided that the pressure remains constant. Mercury-in-glass and alcohol-in-glass thermometers, made by Fahrenheit during the period from 1706 to 1736, were based on three *fixed* temperatures, specified as 0, 32, and 96°, and defined, respectively, as (1) the temperature of a mixture of ice, water, and sal-ammoniac or sea salt, (2) that of a mixture of water and ice, and (3) that of the mouth or armpit of a healthy man. This constituted the original, and very nearly the present, Fahrenheit scale.

1·4 GAS SCALES

Later, hydrogen and nitrogen gas thermometers were instituted. These were based on the Charles' law for ideal gas behavior. Thus

$$T_1/T_2 = p_1/p_2 \tag{1·1}$$

or
$$T_1/T_2 = v_1/v_2 \tag{1·2}$$

where p_1/p_2 represents the ratio of the pressures exerted by a body of gas confined at constant volume at the two temperatures, T_1 and T_2, respectively, or v_1/v_2 represents the ratio of the volumes occupied under constant pressure at these two temperatures.

Such scales depended on the particular material used and did not provide satisfactory means for the precise definition of a fundamental quantity.

1·5 THERMODYNAMIC TEMPERATURE

A precise recognition of the nature of the quantity known as *temperature* is necessary to reliable measurement technique.

The universally accepted *thermodynamic temperature scale* is based on the concept of the ideal reversible Carnot cycle. Here the ratio of two absolute temperatures is defined as that of the heat absorbed to the heat discharged in such a reversible cycle. Thus

$$T_1/T_2 = Q_1/Q_2 \tag{1·3}$$

where Q_1 represents the heat absorbed at the upper temperature T_1, and Q_2, that discharged at the lower temperature T_2. This ratio, and the definition of temperature based upon it, is shown to be independent of the working substance. Although such an ideal cycle is impossible of realization in an actual device, corrections can be calculated to convert the readings on a gas scale to this thermodynamic scale, using thermodynamic theory and the measured properties of the particular gas.[1,2]

1·6 LIMITATION TO EQUILIBRIUM CONDITIONS

As founded on the idea of a reversible process, the definition of the thermodynamic scale is limited in its application to equilibrium situations. Similarly, as based on thermodynamic laws, it is limited in meaning to aggregates of matter containing a sufficient number of elementary particles for statistical uniformity. Thus, the term "temperature" is not to be used in relation either to empty space or to individual electrons, atoms, molecules, etc.

1·7 NONEQUILIBRIUM CONDITIONS

Engineering problems rarely deal with individual particles or empty space. Conditions of thermodynamic equilibrium are, however, more the exception than the rule. Temperature measurements are more often of interest where nonuniformity exists, with corresponding heat-flow patterns, than in situations where an entire region is at one constant, uniform temperature. Similarly, relative motion between solids, between solids and fluids, and turbulent conditions within fluids, particularly at supersonic velocities in gases,[3] are examples of nonequilibrium conditions. A chemical reaction in progress, which may be between groups of constituents within the body of a fluid, between a fluid and a solid at the interface, or between constituents of a fluid on the surface of a bounding solid, provides a nonequilibrium situation. Electrical discharges in gases;[4] ionic currents in liquids; steady and oscillating electric currents in solids; nuclear reactions; and radiations of cosmic rays, neutrons, and ordinary sunshine are additional circumstances where equilibrium is violated.

1·8 THE BASIC COMPLEXITIES OF TEMPERATURE MEASUREMENT

Thus, the facts are (1) that a primary-standard instrument, i.e., an actual device operating on the reversible Carnot cycle, is impossible of realization, and, (2) that most of the conditions in which engineering measurements are required, i.e., nonequilibrium conditions, are cases where the universally accepted definition of temperature fails to apply. These two difficulties constitute a theoretical complexity in the measurement of temperature. This is a type of difficulty not common in the measurement of other quantities.

1·9 TEMPERATURE AT A POINT

Where macroscopic flow of heat or gross mechanical motion is the only violation of equilibrium, it is usually possible to focus attention on local zones large enough for internal statistical uniformity, but small enough so that thermal uniformity may be said to prevail for finite intervals in time. Then, within such small finite intervals in space and time, equilibrium may be said to prevail and the definition of thermodynamic temperature to apply. This gives rise to the concept of *temperature at a point*, which is the thermodynamic temperature of such a zone at a point in space and time within the interval.

The concept of temperature at a point is adequate for dealing with most of the nonequilibrium circumstances encountered in the measurement of internal temperatures in solids.

1·10 PSEUDOTEMPERATURES

In various of the other varieties of nonequilibria, particularly in the gaseous states, disruption of the normal statistical pattern may occur on a microscopic level, i.e., affecting relations among the elementary particles. Here the definition of thermodynamic temperature breaks down fundamentally. It is necessary to set up various other definitions of temperature, usually called *pseudotemperatures*, suited to the individual circumstances. Since such definitions are designed realistically to provide means of measuring that aspect of practical interest in the given situation, they are taken up individually in direct relation to the analyses of these special measurement problems.

All such definitions of pseudotemperatures are so phrased as to refer to the thermodynamic temperature of some related set of conditions. Hence, subject to specialized interpretation, the thermodynamic temperature becomes the absolute standard of reference for nonequilibrium, as well as for equilibrium conditions.[5]

1·11 MEASUREMENT PROCEDURES

The problem of referring measurements by means of actual instruments, such as are suitable in various engineering conditions, to the thermodynamic scale is dealt with through the operational procedure known as the *International Temperature Scale*. This procedure is given therein only for temperatures above the oxygen point. Various *provisional scales* of local acceptance must be resorted to for measuring lower temperatures. The field of very low temperatures is a subject involving fundamental as well as practical questions best discussed in relation to actual problems.[6]

1·12 UNITS OF MEASUREMENT

By definition, the *thermodynamic centigrade*, or *Celsius, scale*, has its zero at the temperature of equilibrium between ice and air-saturated water, and its 100° mark at the temperature of pure water boiling under standard atmospheric pressure. The subdivision of the scale is the same as that of the Kelvin scale, where the zero is at *absolute* zero and on which the freezing point of water has been determined as being approximately 273.16°K,* the boiling point being 100° higher. These are thermodynamic scales.

* The value 273.16°K corresponds to practice in the United States, although the value 273.15°K was recommended by the Ninth General Conference on Weights and Measures.[7]

Sec. 1·13   THE INTERNATIONAL TEMPERATURE SCALE

The *Fahrenheit scale* is defined in terms of the thermodynamic centigrade, or Celsius, scale. Thus, by definition

$$°F = \tfrac{9}{5}°C + 32 \tag{1·4}$$

where °F means degrees Fahrenheit and °C means degrees Celsius (or centigrade). In American and British industrial practice the Fahrenheit scale is usually used.

The *Rankine scale* is a scale for which the zero is intended to be approximately the absolute zero.

$$°R = °F + 459.69 * \tag{1·5}$$

where °R means degrees Rankine. This scale is used in American and British industrial practice where calculations are to be performed in terms of absolute temperatures.

1·13   THE INTERNATIONAL TEMPERATURE SCALE

The procedure here consists in transferring the thermodynamic scale to a series of fixed points by means of the gas thermometer. Precision-type laboratory instruments then serve to interpolate between the fixed points. More convenient types of working-standard instruments may then be calibrated at the *standards* laboratories and *certified* within stated limits of error.[8,9] Quality control in the manufacture of instruments, or day-to-day recalibration of shop or laboratory measuring instruments may then depend on these certified units.[10]

This sequence of instruments, although somewhat complicated, is essentially unavoidable. The gas thermometer is reproducible as being, in principle, dependent only on the properties of a gas, as compared to devices depending on the complex and erratic characteristics of solids. Its readings are transferable to the thermodynamic scale by a relatively simple calculation based on those aspects of thermodynamic and gas theory considered best established.[1,2] In its precise form it is not only an unwieldy device, but, insusceptible of exactness in individual readings. It can, only with the greatest difficulty, be used at elevated temperatures. Similarly, the precision laboratory-standard instruments are not always those best adapted to practical industrial situations.[10]

DEFINITION OF THE INTERNATIONAL TEMPERATURE SCALE OF 1948

1. Temperatures on the International Temperature Scale of 1948 will be designated as "°C" or "°C (Int. 1948)" and denoted by the symbol, t.

* If the value 273.15°K is accepted for the temperature of the ice point, this figure becomes 459.67.[7]

2. The scale is based upon a number of fixed and reproducible equilibrium temperatures (fixed points) to which numerical values are assigned, and upon specified formulas for the relations between temperature and the indications of the instruments calibrated at these fixed points.

3. The fixed points and the numerical values assigned to them are given in Table I. These values, in each case, define the equilibrium temperature

TABLE I. *Fundamental and primary fixed points under the standard pressure of 1,013,250 dynes/cm²*

	Temperature, °C
(a) Temperature of equilibrium between liquid oxygen and its vapor (oxygen point)	−182.970
(b) Temperature of equilibrium between ice and air-saturated water (ice point) (*Fundamental fixed point*)	0
(c) Temperature of equilibrium between liquid water and its vapor (steam point) (*Fundamental fixed point*)	100
(d) Temperature of equilibrium between liquid sulfur and its vapor (sulfur point)	444.600
(e) Temperature of equilibrium between solid and liquid silver (silver point)	960.8
(f) Temperature of equilibrium between solid and liquid gold (gold point)	1063.0

corresponding to a pressure of 1 standard atmosphere, defined as 1,013,250 dynes/cm². The last decimal place given for each of the values of the primary fixed points only represents the degree of reproducibility of that fixed point.

4. The means available for interpolation lead to a division of the scale into four parts.

(a) From 0°C to the freezing point of antimony the temperature, t, is defined by the formula

$$R_t = R_0(1 + At + Bt^2)$$

where R_t is the resistance, at temperature, t, of the platinum resistor between the branch points formed by the junctions of the current and potential leads of a standard resistance thermometer. The constant, R_0, is the resistance at 0°C, and the constants, A and B, are to be determined from measured values of R_t at the steam and sulfur points. The platinum in a standard resistance thermometer shall be annealed, and of such purity that R_{100}/R_0 is greater than 1.3910.

(b) From the oxygen point to 0°C, the temperature, t, is defined by the formula

$$R_t = R_0[1 + At + Bt^2 + C(t - 100)t^3]$$

where R_t, R_0, A, and B are determined in the same manner as in (a) above, and the constant, C, is calculated from the measured value of R_t at the oxygen point.

(c) From the freezing point of antimony to the gold point, the temperature, t, is defined by the formula

$$E = a + bt + ct^2$$

where E is the electromotive force of a standard thermocouple of platinum and platinum-rhodium alloy when one junction is at 0°C and the other is at the temperature, t. The constants, a, b, and c, are to be calculated from measured values of E at the freezing point of antimony and at the silver and gold points. The antimony used in determining these constants shall be such that its freezing temperature, determined with a standard resistance thermometer, is not lower than 630.3°C. Alternatively, the thermocouple may be calibrated by direct comparison with a standard resistance thermometer in a bath at any uniform temperature between 630.3° and 630.7°C.

The platinum wire of the standard thermocouple shall be annealed and of such purity that the ratio R_{100}/R_0 is greater than 1.3910. The alloy wire shall consist nominally of 90 per cent platinum and 10 per cent rhodium by weight. When one junction is at 0°C, and the other at the freezing point of antimony (630.5°C), silver, or gold, the completed thermocouple shall have electromotive forces, in microvolts, such that

$$E_{Au} = 10{,}300 \pm 50\,\mu v$$

$$E_{Au} - E_{Ag} = 1185 + 0.158(E_{Au} - 10{,}310) \pm 3\,\mu v$$

$$E_{Au} - E_{Sb} = 4776 + 0.631(E_{Au} - 10{,}310) \pm 5\,\mu v$$

(d) Above the gold point the temperature, t, is defined by the formula

$$\frac{J_t}{J_{Au}} = \frac{e^{c_2/\lambda(t_{Au}+T_0)} - 1}{e^{c_2/\lambda(t+T_0)} - 1}$$

in which J_t and J_{Au} are the radiant energies per unit wavelength interval at wavelength λ, emitted per unit time by unit area of a black body at the temperature t, and at the gold point, t_{Au}, respectively.

c_2 is 1.438 cm degrees.
T_0 is the temperature of the ice point in °K.
λ is a wavelength of the visible spectrum.
e is the base of Naperian logarithms.*[7]

1·14 RANGE COVERED BY THE INTERNATIONAL TEMPERATURE SCALE OF 1948

Upper limits of the order of 10^{12}°K have been suggested for the phenomenon of temperature. Any such limit would be beyond the range of most practical interest in measurement. No upper limit is

---

* "The temperature 0°C may be realized experimentally well enough for nearly all purposes by the use of a mixture of finely divided ice and water saturated with air at 0°C in a well-insulated container such as a Dewar flask. It is recommended, however, that for work of the highest precision the zero point be realized by means of the triple point of water, a point to which the temperature 0.0100°C has been assigned." Text of The International Temperature Scale of 1948, Part III, Recommendations. (*By permission, Stimson in J. Research Natl. Bur. Standards,* **42**, no. 3, pp. 211–213, March, 1949).[7]

assigned to the scope of the International Temperature Scale. A lower limit for this scale, however, is at the oxygen point, which is 90.19°K.

1·15 PROVISIONAL TEMPERATURE SCALE

The National Bureau of Standards (Washington, D. C.) maintains a *Provisional Temperature Scale* covering the range 11 to 90.19°K. This scale has been based upon a group of resistance thermometers calibrated by direct comparison with a constant-volume helium gas thermometer. Similar scales are maintained at a number of other laboratories, including the Physikalisch-Technische Reichsanstalt (Berlin, Germany), the Kamerlingh-Onnes Laboratory (Leiden, Holland), and the University of California (Los Angeles, California). These scales differ by a few hundredths of a degree. Continuing research is leading to an exact and universal definition of a scale in this range in terms of fixed points for pure materials.[11]

1·16 "INTERNATIONAL" VAPOR-PRESSURE SCALE

At an informal meeting between representatives of cryogenic laboratories in Holland, the United States, and Great Britain, held in Amsterdam, Holland, in July, 1948, a helium vapor-pressure scale was agreed upon. This scale was defined in terms of a series of algebraic relations, empirical coefficients, and correction curves for temperatures above 0°K. Tables computed from these relations were given for helium vapor pressure vs. temperature. The range covered in these tables is from 0.657°K, at which the vapor pressure of helium is given as 0.001 mm mercury, to 5.20°K, at which the helium vapor pressure is 1.720 mm mercury.[6]

1·17 THERMODYNAMIC SCALE BELOW 11°K

Below 11°K and down to about 1°K, the constant-volume helium gas thermometer continues to represent a means of reproducing the thermodynamic scale. The working instruments that function in this range can be referred to this gas scale; or the "International" Vapor-Pressure Scale can be used.

At 1°K, however, the small magnitudes of pressures, the departure of helium from ideal behavior, and adsorption effects have become such that a helium scale is awkward to use. In this region below 1°K, the only parameter that behaves regularly and that can still be readily measured is the magnetic susceptibility of paramagnetic salts. The thermodynamic scale must here be realized in terms of the heat-entropy-temperature relations occurring in the process of *magnetic cooling.* Thus

$$T = \Delta Q/\Delta S \tag{1.6}$$

where ΔQ represents the heat removed during isothermal magnetization, ΔS, the resulting entropy decrease, and T, the absolute temperature.[6,12]

1·18 MAGNETIC SCALE

In the region below 1°K, the approximate Curie law becomes the most convenient basis for temperature measurement. Thus

$$I = cH/T_m \tag{1.7}$$

where I represents the intensity of magnetization produced in the sample of paramagnetic salt by the imposed magnetic field of strength H at the absolute temperature T_m of the salt. T_m is the temperature on the *magnetic scale*. c is a constant which can be determined by calibration against the helium gas thermometer or the helium vapor-pressure scale at temperatures above 1°K. Since the indicated magnetic moment depends on the shape of the specimen, the usual practice is to correct to that which would be displayed by a spherical sample. As with the gas scale, corrections can be computed in terms of thermodynamic theory and the measured properties of a particular salt to refer readings on the magnetic scale for that salt to the thermodynamic scale.

At sufficiently low temperature, departures of this phenomenon from ideal behavior and corresponding discrepancies between the magnetic scale and the absolute thermodynamic scale become substantial. Temperatures as low as 0.003°K have been identified on the thermodynamic scale, whereas 0.001°K has presumably been attained, and 0.0001°K predicted.[12]

1·19 SHOP STANDARDS

The International Temperature Scale can be set up autonomously in any suitably equipped laboratory. The instruments defined are precise, reproducible, and suitable for making actual temperature measurements.[7] Consequently, much precise scientific work is done directly on this scale. The precision attained is, however, frequently unnecessary, even in the locally prevailing primary-standard instruments. Moreover, arranging to reproduce the International Temperature Scale may be inconvenient or altogether unfeasible.[10]

The various standards laboratories, such as the National Bureau of

Standards (Washington, D. C.), maintain means of reproducing the International Temperature Scale together with arrangements for making comparisons with other instruments sent to them for calibration.[8, 9] Whereas this service can be obtained on any of a number of types of instrument, the relatively cheap, compact, self-contained character of the mercury-in-glass thermometer renders this instrument peculiarly convenient for this purpose. Thus, certified mercury-in-glass thermometers are widely used as local primary, secondary, and subsecondary standards.[10] The local problems of standardization then depend upon the behavior characteristics of mercury-in-glass thermometers.

1·20 ERRORS IN MERCURY-IN-GLASS THERMOMETERS

The difficulties in the precision use of the mercury-in-glass thermometers arise through the fact that glass, being a supercooled liquid, is subject to viscous flow under stress, even at room temperature. The stresses causing such flow may be temperature stresses incurred in fabrication or in use at elevated temperatures. This viscous flow causes progressive change in the bulb volume and, hence, in the calibration.[13]

Progressive changes in thermometers made from good grades of thermometer glasses do not exceed 0.2°F, provided that the instrument has not been heated above 300°F. If accuracy greater than to within 0.2°F is required, the ice point must be checked as part of the job of taking a reading. If the thermometer is to be used at temperatures higher than 300°F, i.e., up to 900°F, the progressive changes cannot be predicted but can be expected to amount to more than 0.2°F. If the thermometer is not used above 750°F, the change can be assumed to be about the same as that of the ice point. From 750 to 950°F and above, there is likely to be some change in the stem as well and correction at the ice point is not sufficient.[13]

For good grades of thermometer glass the *hysteresis* effect, due to heating and cooling again, is less than 0.01°F for each 10°F that the bulb is heated above the original temperature, provided that the thermometer is not heated above 300°F. This depression, due to heating and cooling, disappears almost entirely in a few days' time. If the thermometer is heated above 300°F, the changes become erratic and cannot be predicted. To avoid error due to hysteresis, one must take readings in order (after checking the ice point), starting with the lowest temperature and proceeding to the higher readings.[13]

The effect of external pressure on the bulb is to make the thermometer tend to read high. For bulb diameters of around ¼ in., this amounts to about 0.2°F per atmosphere.[13]

If a thermometer is calibrated for total (or partial) immersion, it must be used only in that way. Similarly, if a thermometer is calibrated for horizontal (or vertical) use, it must be used only in that way, or a correction must be applied.

Thus, if the mercury thread of a thermometer is not at the same temperature as the bulb, the emergent-stem correction Δt, °F, to be added algebraically to the indicated temperature, can be computed from

$$\Delta t = aN(t_b - t_a) \tag{1·8}$$

where a is 0.00009; N is the number of degrees on the scale by which the mercury thread is emergent from the bath; t_b is the temperature indicated by the thermometer, °F; and t_a is the mean temperature of the emergent stem, °F.[13–16]

If a thermometer has not been properly annealed and is heated to, say, 600°F, it may progressively anneal at a rapid rate at that temperature. The resulting change in calibration may be as much as 35°F.[13]

For ¼-in. diameter bulbs suddenly immersed in a well-stirred water bath, ¼ to ½ min is required before equilibrium is reached. In still air, about 10 min may be required. For larger bulbs, the time required would be proportionately longer.[13, 15, 17–19]

1·21 FIXED POINTS

Locally reproducible fixed points are widely used for shop and laboratory standardization of instruments.[14] The melting point of ice is the most commonly used fixed point. The vapor above water, boiling in a *hypsometer* at a measured barometric pressure, is used with tables giving the vapor pressures at various temperatures. A large assortment of other pure substances is available for this purpose.[20] Perhaps the most convenient and precise arrangement is the *benzoic acid cell*, as furnished by the National Bureau of Standards (Washington, D. C.) in a self-contained unit complete with a well for the insertion of a thermometer bulb. This yields a temperature of 122.36°C (252.25°F) with the accuracy stated to the nearest 0.003°C (0.0054°F).[10]

REFERENCES

1. F. G. Keyes, "Gas Thermometer Scale Corrections Based on an Objective Correlation of Available Data for Hydrogen, Helium, and Nitrogen," American Institute of Physics, *Temperature*, pp. 45–59, Reinhold Publishing Corp., New York, 1941.
2. J. R. Roebuck and T. A. Murrell, "The Kelvin Scale from the Gas Scales by Use of Joule-Thomson Data," American Institute of Physics, *Temperature*, pp. 60–73, Reinhold Publishing Corp., New York, 1941.

3. K. F. Herzfeld, "Relaxation Effects in Shock Waves," *Symposium on Aero-Thermodynamics*, NOLR 1134, pp. 57–67, U. S. Naval Ordnance Laboratory, White Oak, Silver Spring (1950).
4. F. L. Mohler, "Concepts of Temperature in Electric Discharge Phenomena," American Institute of Physics, *Temperature*, pp. 734–744, Reinhold Publishing Corp., New York, 1941.
5. H. D. Baker and G. L. Laserson, "An Investigation Into the Importance of Chemiluminescent Radiation in Internal Combustion Engines," *General Discussion on Heat Transfer*, London Conference, Sec. IV, 7 pp., Institution of Mechanical Engineers, London, and American Society of Mechanical Engineers, New York (1951).
6. C. T. Lindner, "The Measurement of Low Temperatures," *Research Report* R-94433-2-A, pp. 1–20, 29–36, Westinghouse Research Laboratories, East Pittsburgh (1950).
7. H. F. Stimson, "The International Temperature Scale of 1948," RP 1962, *J. Research Natl. Bur. Standards*, 42, no. 3, pp. 209–217 (March, 1949).
8. "Testing of Thermometers," *Natl. Bur. Standards Circ.* 8, 4th edition, 18 pp., Government Printing Office, Washington (1926).
9. "Testing by the National Bureau of Standards," *Natl. Bur. Standards Circ.* 483, pp. 52–55, Government Printing Office, Washington (1949).
10. W. D. Wood, "The Development and Maintenance of Calibrating Standards by an Instrument Manufacturer," *Instruments*, 22, no. 11, pp. 1004–1007 (November, 1949).
11. H. J. Hoge, "Vapor Pressure and Fixed Points of Oxygen and Heat Capacity in the Critical Region," RP 2081, *J. Research Natl. Bur. of Standards*, 44, no. 3, pp. 321–322, 343–345 (March, 1950).
12. C. F. Squire, "Magnetic Cooling; Production and Measurement of Temperatures Below 1°K," American Institute of Physics, *Temperature*, pp. 745–756, Reinhold Publishing Corp., New York, 1941.
13. J. Busse, "Liquid-in-Glass Thermometers," American Institute of Physics, *Temperature*, pp. 228–255, Reinhold Publishing Corp., New York, 1941.
14. P. D. Foote, C. O. Fairchild, and T. R. Harrison, "Pyrometric Practice," *Technological Papers of the Bureau of Standards* 170, pp. 11–12, 14–17, 225–282, Government Printing Office, Washington (1921).
15. American Society of Mechanical Engineers, "Liquid-in-Glass Thermometers," *Power Test Codes Series 1929*, "Instruments and Apparatus," Part 3, "Temperature Measurement," Ch. 6, pp. 16–26, New York (1931).
16. C. E. Arreger, "Precautions in Using Thermometers," *Can. Chem. Process Ind.*, no. 1289 (April, 1945).
17. J. G. Durham, "A Note on the Care of Liquid-in-glass Thermometers," *J. Sci. Instr. and Phys. in Ind.*, 26, no. 6, pp. 205–206 (June, 1949).
18. V. Hiergesell, "Precision Laboratory Thermometers, Their Construction, Production and Uses," *Instruments*, 22, no. 9, pp. 802–803 (September, 1949).
19. The American Society for Testing Materials, "Method of Testing and Standardization of Etched-Stem Liquid-in-Glass Thermometers" (Tentative), *ASTM Standards on Thermometers*, E77-49, pp. 322–340, Philadelphia (1951).
20. W. F. Roeser and H. T. Wensel, "Methods of Testing Thermocouple Materials," American Institute of Physics, *Temperature*, pp. 289–291, Reinhold Publishing Corp., New York, 1941.

2

METHODS FOR MEASURING TEMPERATURE

2·1 CLASSIFICATION

Actual devices available for making temperature measurements can be classified into three groups: (1), where the body, whose temperature is to be measured, serves as its own thermometer; (2), where a solid thermometer element is inserted into the body; and, (3), where the temperature-measuring device or *pyrometer* operates at a distance by radiation emanating from the body.

2·2 BODY ITS OWN THERMOMETER

The pressure of a body of confined gas, or its volume at a constant pressure, can be used to indicate its mean temperature. Similarly, the pressure exerted by the vapor of a substance in equilibrium with the solid or liquid phase can be used. The velocity of sound in a body of gas depends on temperature, and its measurement will serve to determine the mean temperature of this gas along the path of the sound wave. The electrical resistance of a solid body is a function of temperature [1] and is a convenient means for measuring the mean temperature of such components as heater coils. Permanent or temporary changes in the hardness or other properties of solids may be characteristic of the temperatures to which they have been raised under given conditions. The melting and boiling points are familiar indications. Solids and liquids expand thermally, and such thermal expansion can be used to indicate a mean temperature in the portion of the body whose expansion is measured. A thermoelectric electromotive force, or *emf*, is used to indicate an average interface temperature, as, for example, in the functioning of cutting tools. The *Johnson noise*, caused by the random motions of the electrons in solid bodies, has been made to serve as an indicator of their absolute temperatures.[2] Similarly, the magnetic susceptibilities of certain materials, usually at very low temperatures, are used to indicate temperatures. Characteristics of organic growths may reflect the temperature conditions under which these have developed. In general, any property having an appreciable and consistent rate of temperature variation will serve to indicate temperature.

2·3 THERMOMETERS

What we usually call thermometers are bodies devised to indicate their own temperatures and so proportioned as to permit their insertion into other bodies whose temperature is desired to be measured. The term "pyrometer" is often applied to instruments that operate at temperatures other than the temperature to be measured. Thermometers may function in accordance with any of the principles listed in Sec. 2·2. Indicating instrumentation may be self-contained or external, i.e., at a distance from the *sensitive element*. The primary condition involved in the use of such thermometers is the presumed existence of thermal equilibrium between the sensitive element of the thermometer and the adjacent *parent-body material*. It must also be supposed that the presence of the thermometer does not in itself excessively alter the temperature to be measured.

2·4 RADIATION PYROMETERS

In the *radiation pyrometer* the sensitive element does not achieve thermal equilibrium with the portion of the body whose temperature is to be measured. The sensitive element, here, responds to the quantity or quality of the thermal radiation emitted by such a portion, which must thereby be a spot on the surface. The quantity may be a calibrated fraction of the total emitted radiation, the intensity thereof within a given *wavelength* range, or the relative intensities in two or more selected wavelengths. Indicating instrumentation is always at a distance from the radiating body. This distance may be large, as with temperature measurements of the heavenly bodies. Indicating instrumentation may be self-contained or at a distance from the sensitive element. The fact that no portion of the measuring device is required to assume the temperature of the body under investigation renders this method peculiarly adaptable to the measurement of elevated temperatures.[3]

2·5 GAS THERMOMETERS

Thermometers based on the known relationships between pressure, specific volume, and temperature for various gases are called *gas thermometers*. These occur both as bulky and awkward primary-standard instruments, and in rugged commercial devices known as *Class-III systems*.*

* Classification: *Bourdon Thermometer* (*liquid*), Class I (see Sec. 2·7); *Vapor-Pressure Thermometer*, Class II (see Sec. 2·6); *Gas Thermometer*, Class III (see Sec. 2·5); and *Thermocouple*, Class IV (see Sec. 2·10).

Sec. 2·8 ELECTRIC-RESISTANCE THERMOMETERS 15

In the latter form they permit indication at distances as great as 200 ft or more and are used in the temperature range of -60 to $1000°F$, or higher, with accuracies of ± 1 per cent and response to changes as small as $0.01°F$. The sensitive element, i.e., the *bulb*, may be as small as $\frac{3}{4}$ in. diameter. Because a Bourdon-type pressure gage is often used for indication, these are commonly called *Bourdon thermometers*.[3-8]

2·6 VAPOR-PRESSURE THERMOMETERS

Vapor-pressure thermometers are similar in construction and operating features to gas thermometers. As a Bourdon gage is often used for indication in the commercial forms, they also go by the name, Bourdon thermometers, and are distinguished specifically as *Class-II systems*. The pressure of the vapor in coexistence with the liquid phase, being a function of temperature independent of specific volume, is the working principle. With various fluids, temperatures from -20 to $700°F$ can be measured; however, the working range for any one liquid is limited to 200 or $300°F$. Vapor-pressure thermometers are used in precision form as primary-standard instruments in the low-temperature region above $1°K$ (see Sec. 1·16).[3,4,5,7]

2·7 LIQUID-FILLED THERMOMETERS

Liquid-filled thermometers are available in diverse forms, as, for example, mercury-in-glass thermometers. These are usually compact, self-contained, and direct-reading instruments; but they also occur with Bourdon-gage indication at a distance, similarly to commercial gas and vapor-pressure thermometers, in which form they are designated as *Class-I systems*. With various liquids, ranging from alcohol to gallium, temperatures from -170 to $2200°F$ can be measured, whereas the Beckmann type is sensitive to changes as small as $0.01°F$. However, the scope of the range is in inverse ratio to the sensitivity, and any one instrument is likely to be of narrowly limited range.[3,4,5,7,9-13]

2·8 ELECTRIC-RESISTANCE THERMOMETERS

The *electric-resistance thermometer* usually consists of a small coil of wire, with auxiliary instruments to measure its electrical resistance. This resistance depends on the temperature of the coil and thus serves to yield measurements of temperature. Although this instrument is regarded as the most precise and reliable for the range -297.35 to $1166.9°F$, attainment of this precision requires much care in construction and use. This thermometer is also made commercially in various convenient and rugged forms of lower precision. The minimum coil diameter is usually ¼ in. Otherwise, in order to measure a space average of temperature, wire can be extended to traverse a large region. Indication can be arranged at a distance with unusual convenience, because wires that are not sensitive to thermal or mechanical disturbances are used to effect the connection.[3,5,7,14] Certain *electrolytic*[15] and nonmetallic solid resistors, called *thermistors*,[1] provide large response coefficients with corresponding loss in reliability.

At the point of transition to the superconducting state in low-temperature phenomena, the electrical resistance of pure metals drops discontinuously to zero. For attainable degrees of purity, the transition curves are steep, with correspondingly high temperature coefficients of resistance within these narrow temperature ranges in the neighborhood of 0°K. Radiation detectors, based on resistance elements operating in such ranges, are extremely sensitive.

2·9 BIMETALLIC THERMOMETERS

A *bimetallic* strip, consisting of autogenously welded layers of materials of differing coefficients of thermal expansion such that it warps with changing temperature, serves as the sensitive element. Depending thus entirely on an elastic property, this kind of element is not suited to high precision. The positive mechanical motion provided is convenient for actuating automatic recorders, controls, and telemetering devices. All-metal, rugged, easy-reading, self-contained instruments are available in sufficiently compact forms to compete with the mercury-in-glass type in this respect. Bimetallic thermometers can be used at temperatures up to 500°F (or 1000°F for special constructions) yielding accuracies of ±1 to ±5 per cent. Plastic bilayers can be made with tenfold greater sensitivity, but are subject to more severe limitations otherwise.[3,5,7,16,17]

2·10 THERMOCOUPLES

The *thermocouple*, or *thermoelectric thermometer*, also designated as a *Class-IV system*, is probably the most versatile of temperature-measuring instruments. It has been applied over the entire range from the immediate vicinity of absolute zero to 5400°F,[18] whereas in the range 1166.9 to 1945.4°F it is specified in the International Temperature Scale as the most precise and reliable of all instruments. The temperature-sensitive element in a thermocouple can be made arbitrarily small, thus facilitating precise measurement of temperature at a point. The low thermal capacity of the element, resulting from this small size, tends toward quick response and facilitates measurement of tempera-

Sec. 2·11 RADIATION AND OPTICAL PYROMETERS

ture at an instant. The thermocouple itself is very cheap and simple in construction. The indicating instruments are usually quite elaborate. They, however, can be located at a distance from the body and used for reading a large number of thermocouples alternately. The thermoelectric thermometer may be in the form of a very rugged instrument of low precision, in the form of a highly refined instrument, or in any intermediate form.[3,5,7,19]

2·11 RADIATION AND OPTICAL PYROMETERS

The *radiation pyrometer* consists of a blackened disk suitably shielded from stray radiation and provided with sensitive means to indicate small changes in its surface temperature. Exposed to the "view" of a "hot" body, this disk is warmed by the emanating radiation and assumes a temperature much lower than, but related to, that of the portion of the surface of the body to which it is exposed. The readings on the temperature-indicating device, which may be at a distance, are calibrated under operating conditions to read an *apparent* corresponding temperature on the "hot" body. A calculated correction is required, based on an assumed value for the mean emissivity of the "hot" body, to convert this apparent temperature to the indication of actual surface temperature. If other radiating bodies are in the vicinity, further calculated corrections are required to account for reflected radiation originating in these other bodies.[3,5,7,20,21]

The entire spectrum is often employed, giving rise to the name *total-radiation pyrometer*. The relative intensities in two fractions of the spectrum can, however, be used to realize the *two-color principle*. The ratio of two such intensities provides a measure of temperature that does not require correction for emissivity, provided that the mean emissivities can be assumed to be the same in both wavelength ranges. Corrections are still required for the effects of other radiating bodies in the vicinity.[21,22]

The *optical pyrometer* consists of a special portable telescope, through which a portion of the surface of the body is viewed. An electrically heated glowing filament is located in the image plane of this telescope. The temperature of this filament can be adjusted over a certain range, the various settings being indicated on a dial. The image and filament are viewed together through the eyepiece of the telescope which is provided with a *filter*, usually a disk of dyed glass, transparent in but a very narrow range of wavelengths of light. To read the temperature of a spot on the surface of the "hot" body, the telescope is focused on that spot while the filament current is adjusted until, as viewed in the given wavelength, the body and glowing filament appear equally bright. At this point the filament seems to *disappear* against its background. The dial is usually calibrated to read apparent temperature directly. If the "hot" body is at a higher temperature than any to which it is feasible to adjust the electric filament, a filter is used before the image to reduce its brightness to within the range attainable by the filament. The presence of this filter changes the significance of the dial readings. Either a conversion factor or a second direct-reading scale is provided.[3, 5, 7, 21, 23, 24]

The *two-color optical pyrometer* depends on the ratio of the intensities in two narrow wavelength bands.

Neither the radiation nor the optical type is subject to an upper limit in the temperature range to which it is adapted; however, application becomes difficult at temperatures below 1350°F for the optical and 300°F for the radiation-type instruments. Whereas the optical pyrometer is specified in the International Temperature Scale for temperatures above 1945.4°F, and although an unskilled operator can repeat readings consistently to within ±10°F on this instrument,[25] both the radiation and optical pyrometers must be used with a great deal of discretion if any significance is to be attached to their readings. For example, a bank of freshly fallen snow viewed in daylight may read a temperature of several thousand degrees Fahrenheit. It is difficult and frequently impossible to make the required corrections with any degree of accuracy. Errors of several hundred degrees are not uncommon.[21, 24]

2·12 PYROMETRIC CONES

Pyrometric cones are slender trihedral pyramids made of such mixtures of minerals that when heated uniformly at a specified rate they deform on reaching temperatures characteristic of these mixtures. They are available commercially in a series of 61 cone numbers approximately equally spaced over the temperature range from 1085 to 3659°F. As used in the ceramic industry they are not, however, regarded as temperature-measuring devices, but rather as indicators of the cumulative effects of heat treatment in similar, adjacent, ceramic materials during "firing." [3, 26-28]

2·13 CRAYONS, PELLETS, PAINTS

Mineral mixtures of definite melting temperatures are available commercially in the forms of *crayons*, *paints*, and *pellets*. The temperatures, in the range from 113 to 2500°F, corresponding to the particular numbers in the series, are indicated by the advent of a "wet" or molten

Sec. 2·17　RADIATION AND DENSITY MEASUREMENTS　19

appearance. As applied to the surface of a body, they indicate by this "wet" appearance when the local surface area has reached the corresponding temperature.[29, 30]

2·14　COLOR INDICATORS

In the heat treatment of steel, tempering temperatures have long been estimated by the colors attained by surface oxide formations as these develop during the heating. A special alloy has been produced to indicate temperature by its vividly changing color in the range 932 to 1620°F. Ceramic color indicators in the forms of paints and capsules are also available.[29, 30]

2·15　INFRARED PHOTOGRAPHY

In the temperature range of from 600 to 850°F, infrared photographs may record surface-temperature distributions. The more "luminous," hotter areas thereby appear as the "highlights," whereas the cooler areas are darker on the print.[31-33]

Prior coating with certain phosphors, whose luminescence is temperature dependent, renders surface thermal patterns in this range visible to the human eye and, thus, also subject to ordinary photography and photoelectric cells as means of temperature measurement.[34]

2·16　KURLBAUM AND FERY LINE-REVERSAL METHODS

Temperatures of semitransparent media, such as gases, can be measured, applying Kirchhoff's law, by viewing an incandescent, *blackbody background* through the medium. The medium may be *colored* by an additive substance to increase its opacity. Its uniform temperature is then that of the background, when the latter's luminosity, thus viewed, is unaltered. Observation may be with the human eye or by means of instrumentation, and over the entire spectrum or in a selected wavelength range.

The significance of such measurements, however, as applied to media in states of thermodynamic nonequilibrium, whether of thermal or chemical character, is dubious.[35-37]

2·17　RADIATION AND DENSITY MEASUREMENTS

The relative or absolute intensities of radiation in the various wavelengths from a gas can serve to determine its temperature. The absorptions of beams of X-rays, α-particles, or electrons passed through a layer of the gas, in indicating local gas density, can be used to determine relative or absolute gas temperatures. Similarly, the refractive indexes, determined with an interferometer or by Schlieren photography, are measures of gas densities, and thereby of temperatures.[38]

2·18 THE VELOCITY-OF-SOUND AND ORIFICE METHODS

The velocity of sound in a body of gas, as dependent on the ratio of the adiabatic elasticity to the local density of the gas, is a function of temperature, and is thereby a means for the measurement of temperature. Similarly, the pressure drop across an orifice can be used to determine the temperature of the gaseous flow. These two methods depend on the fundamental laws and properties of gases. In this respect, they are analogous to the primary-standard gas thermometer.[38]

2·19 CONCLUSION

Because of its remarkable combination of characteristics, the thermocouple is considered to be best suited for application in problems of measurement of internal temperatures in solids. Consequently, the remaining chapters of Vol. I will assume the thermoelectric method and deal therewith. Other methods suited to other varieties of problems will be dealt with in Vol. II.

REFERENCES

1. J. A. Becker, C. B. Green, and G. L. Pearson, "Properties and Uses of Thermistors—Thermally Sensitive Resistors," *Electrical Engineering Transactions*, **65**, Trans. 711, pp. 711–725 (November, 1946).
2. J. B. Garrison and A. W. Lawson, "An Absolute Noise Thermometer for High Temperatures and High Pressures," *Rev. Sci. Instr.*, **20**, no. 11, pp. 785–794 (November, 1949).
3. A. G. Worthing and D. Halliday, *Heat*, pp. 19–57, John Wiley & Sons, New York, 1948.
4. American Society of Mechanical Engineers, "Bourdon Tube Thermometers," *Power Test Codes Series 1929*, "Instruments and Apparatus," Part 3, "Temperature Measurement," Ch. 7, pp. 27–35, New York (1931).
5. "Apparatus Directory," *J. Appl. Phys.*, **11**, no. 6, pp. iv–xx (June, 1940).
6. Brown Instrument Company, "Response Speeds of Pressure Type Thermometers," *Bulletin* 60–1, 8 pp., Philadelphia (1942).
7. British Standards Institution, "Temperature Measurement," *British Standard Code*, B.S. 1041: 1943, pp. 10, 16–40, London (1943).
8. E. E. Modes, "Pressure-Temperature Relations in Gas Filled (Class III) Thermometers," *Paper* 50-A-48, 10 pp., American Society of Mechanical Engineers, New York (1950).
9. American Society of Mechanical Engineers, "Liquid-in-Glass Thermometers," *Power Test Codes Series 1929*, "Instruments and Apparatus," Part 3, "Temperature Measurement," Ch. 6, pp. 16–26, New York (1931).

REFERENCES

10. U. S. Treasury Department, Procurement Division, *Federal Standard Stock Catalogue* Sec. IV, Part 5, "Federal Specification for Thermometers; Industrial," GG-T-321, 14 pp., Government Printing Office, Washington (1931).
11. U. S. Treasury Department, Procurement Division, *Federal Standard Stock Catalogue*, Sec. IV, Part 5, "Federal Specification for Thermometers; Clinical," GG-T-311, 8 pp., Government Printing Office, Washington (1933).
12. U. S. Department of Commerce, "Clinical Thermometers, A Recorded Voluntary Standard of the Trade," *Commercial Standard* CS 1-52, 14 pp., Government Printing Office, Washington (1942).
13. American Society for Testing Materials, *Standard Specifications for ASTM Thermometers* E1-51, pp. 278–321, Philadelphia (1951).
14. American Society of Mechanical Engineers, "Resistance Thermometers," *Power Test Codes ASME 1945*, "Supplement on Instruments and Apparatus," PTC 19.3.4—1945, Part 3, "Temperature Measurement," Ch. 4, pp. 5–17, New York (1945).
15. D. N. Craig, "Electrolytic Resistors for Direct-Current Applications in Measuring Temperatures," RP 1126, *J. Research Natl. Bur. Standards*, 21, no. 2, pp. 225–233 (August, 1938).
16. American Gas Association, "A Study of Bimetallic Thermal Elements," *Research Bulletin* 42, 25 pp., New York (1947).
17. H. A. Pohl, "Supersensitive Thermoelement," *Rev. Sci. Instr.*, 22, no. 5, p. 345 (May, 1951).
18. A. Schulze, *Metallische Werkstoffe für Thermoelemente* (Metallic Materials for Thermocouples), pp. 72–79, Edwards Brothers, Ann Arbor, 1946.
19. American Society of Mechanical Engineers, "Thermocouple Thermometers or Pyrometers," *Power Test Codes ASME 1940*, "Information on Instruments and Apparatus," PTC 19.3.3—1940, Part 3, "Temperature Measurement," Ch. 3, pp. 3–21, New York (1940).
20. American Society of Mechanical Engineers, "Radiation Pyrometers," *Power Test Codes ASME 1936*, "Instruments and Apparatus," Part 3, "Temperature Measurement," Ch. 2, pp. 3–10, New York (1936).
21. W. T. Reid and R. C. Corey, "Errors in Temperature Measurement by Radiometric Methods," *Combustion*, 15, no. 8, pp. 30–34 (February, 1944).
22. G. Naeser, "Zur Farbpyrometrie" (Color Pyrometry), *Mitteilungen aus dem Kaiser-Wilhelm-Institute für Eisenforschung zu Düsseldorf*, 12, Lief. 18, Ab. 163, pp. 299–316 (July, 1930).
23. American Society of Mechanical Engineers, "Optical Pyrometers," *Power Test Codes ASME 1940*, "Information on Instruments and Apparatus," Part 3, "Temperature Measurement," Ch. 8, pp. 5–10, New York (1940).
24. M. Jakob, "Balance of Radiation in Employing Optical Pyrometry," *Combustion*, 16, no. 2, pp. 49–50 (August, 1944).
25. W. E. Forsythe, "Optical Pyrometry," American Institute of Physics, *Temperature*, pp. 1127–1129, Reinhold Publishing Corp., New York, 1941.
26. American Society of Mechanical Engineers, "Pyrometric Cones," *Power Test Codes ASME 1929*, "Instruments and Apparatus," Part 3, "Temperature Measurement," Ch. 5, pp. 13–15, New York (1931).
27. G. A. Bole, "Pyrometric Cones," American Institute of Physics, *Temperature*, pp. 988–995, Reinhold Publishing Corp., New York, 1941.

28. Edward Orton Jr. Ceramic Foundation, *The Properties and Uses of Pyrometric Cones*, pp. 9–45, Columbus, 1946.
29. L. C. Tyte, "Temperature Indicating Paints," *Inst. Mech. Engrs. J. and Proc.*, 152, no. 2, pp. 241–243 (September, 1945).
30. K. Guthmann, "Temperaturmessfarben u. Messfarbstifte" (Temperature Measuring Colors and Pencils), *Archiv für technisches Messen*, no. 152, T4-5 (July, 1947).
31. R. Hase, "Untersuchung der Abkühlung glühender Silitstäbe mittels photographische Pyrometrie" (Investigation of the Cooling of Incandescent Silicon Carbide Rods by Means of Photographic Pyrometry), *Z. Tech. Phys.*, 13, pp. 410–415 (August, 1932).
32. M. W. Wallace, *Qualitative Determination of the Heat Distribution on Radio Tubes by Infrared Photography*, Ph.D. Dissertation, Massachusetts Institute of Technology, Cambridge, 1936.
33. W. Clark, *Photography by Infrared: Its Principles and Applications*, pp. 325–327, John Wiley & Sons, New York, 1946.
34. F. Urbach, N. R. Nail, and D. Pearlman, "The Observation of Temperature Distributions and of Thermal Radiation by Means of Non-Linear Phosphors," *J. Opt. Soc. Amer.*, 39, no. 12, pp. 1011–1019 (December, 1949).
35. A. G. Gaydon, *Spectroscopy and Combustion Theory*, pp. 168–175, Chapman & Hall, London, 1948.
36. H. J. Buttner, I. Rosenthal, and W. G. Agnew, "A Study of Flame Temperatures as Determined by the Sodium Line Reversal Method in Totally and Partially Colored Flames," U. S. Navy, Project Squid, *Technical Memorandum* Pur-13, 61 pp., Purdue University, Lafayette (1949).
37. B. Lewis and G. von Elbe, *Combustion Flames and Explosions of Gases*, pp. 258–260, 682–695, Academic Press, New York, 1951.
38. C. G. Suits, "High-Temperature Gas Measurements in Arcs," American Institute of Physics, *Temperature*, pp. 720–733, Reinhold Publishing Corp., New York, 1941.

3

PRECISION REQUIREMENTS

3·1 PLANNING A PROJECT

Selection of a method for any measurement, and design of its mode of application, requires numerical understanding of the required degree of precision. Determination of the proper precision requirements must be made with reference to the larger project for which the data are desired. Instead of relying on guesswork, it is necessary to apply principles on a quantitative basis.

Since industrial projects are usually complex and may have several objectives involving the measurement of various quantities, the direct temperature measurement of one quantity rarely constitutes an entire project. Design of the apparatus is decided in relation to these objectives as a whole, and the available means.

The nature of the design will usually fix one or more factors in such a way as to limit directly the precision with which certain quantities can be measured. This will indirectly limit the precision for other dependent quantities, which are of interest but are not measured independently.

Thus, a decision to work with an operating internal-combustion engine may imply limitations on the constancy with which speed, torque, and rate of fuel consumption can be maintained, and, consequently, on the precision with which engine temperatures can be measured. These, in turn, may limit the precision with which power, efficiency, and various thermal currents can be determined. Such limitations are accepted with the apparatus design itself and will usually have a determining influence on its choice.

With some quantities to be measured, it may be possible, within the limitations of the general design, to achieve greater precision than is of practical interest to the project. The method of measuring other quantities, which are definitely desired, may be independent of the general design. Decision on instrumentation and procedure must then be made after a comparison of estimates of the expected advantages of greater precision with the increased costs in measurement.

Thus, again in the case of an internal-combustion engine, the average rate of heat removal from the system by the circulating oil might be measured by the most refined calorimetric procedure. The advantages accruing from this degree of precision might not, however, warrant the required effort.

After precision standards have been set for the quantities of final interest to the project, a calculation can be made of the precision requirements for the variety of specific measurements from which the final values are deduced. This process should be done in accord with the simple rules described in this chapter.

3·2 ERRORS

Numbers used to describe precision may represent various factors. Thus, temperature is a measure of a condition in a body at a given point and time. Error in measurement may lie in faulty knowledge of the magnitude, of the location, or of the time at which this temperature occurs.

Errors in temperature measurement can be described in several ways.

1. Errors can be stated in absolute units (°F, in., or sec). Absolute units in temperature measurement are defined in terms of the International Temperature Scale with respect to which shop instruments are calibrated.

2. Errors can be stated in relative units. These may be in absolute differences in readings of a given shop instrument or in per cent differences. Thus, differences or ratios between measured quantities may be of greater interest than the quantities themselves.

3. Errors in temperature or other measurements can be stated in terms of their effect on the realization of the work. Thus, x units error in a given temperature measurement may be thought of as contributing y units to the total error in the measurement of some particular heat quantity.

It is a fundamental truism that, since all measurements are merely imperfect attempts to ascertain an exact natural condition, the actual magnitude of a quantity can never be known. Hence, the actual error in the measurement of a quantity is also unknowable. Errors should therefore be discussed only in terms of probabilities, i.e., there may be one probability that a measurement is correct to within a certain specified margin and another probability for correctness to within another specified margin.[1]

3·3 PRECISION

The precision requirement for a measurement can be expressed in terms of the magnitude of error permitted. A magnitude of error can be described in one of two ways: (1), the *probable error* and, (2), the *allowable error*.[1]

The probable error P is that magnitude of error which is most likely to occur. By the definition of the term, probable error, there is an equal likelihood of error larger or smaller than P. Although the probable error defines the precision of the instrumentation as satisfactorily as any alternative criterion, it does not directly indicate the usefulness to be expected in the resulting measurements. The data will probably contain errors larger than the probable error, some of them considerably larger.[1]

Precision is best specified in terms of allowable error A, errors greater than which occur so rarely as to be practically negligible, i.e., occurring, probably, once in the averages of 300 pairs of duplicate measurements. Thus, the confidence that should be placed in the data is commensurate with the error that is known to be quite certainly absent. Since the size and rarity of errors increase simultaneously, the allowable error A is larger than the probable error P, in fact, it is 3 times larger. Thus [1]

$$A = 3P \qquad (3\cdot 1)$$

3·4 SYSTEMATIC AND ACCIDENTAL ERRORS

In terms of their sources (quite apart from means of describing their magnitudes), errors are usually classified as *systematic errors* and *accidental errors*. These terms apply to individual factors contributing to the total error in the measurement of any one quantity.[1]

Systematic errors are those that are inherent in the means of measurement. Thus, if, in some indetectable manner, an instrument always indicated x units "high" (or "low") under a given set of operating conditions, x would be a systematic error. As such, systematic errors are repeated consistently throughout all measurements made under the given conditions; they are not reduced by the taking of additional readings or by the averaging of repeated readings.[1]

Accidental errors are of two kinds, *large* and *small*. Large errors may occur when an operator makes a simple blunder, such as a mistake in arithmetic, failure to close an appropriate switch, or reading from the wrong scale.[1]

Small errors may occur as any of a variety of accidental factors influence individual readings, sometimes this way, sometimes that, with an equal probability in either direction. Thus, an operator, in taking a reading, may estimate the last significant figure. Friction in an instrument may result in the pointer stopping slightly short of its natural equilibrium position, or momentum may carry it beyond.[1]

3·5 TREATMENT OF ACCIDENTAL ERRORS

Accidental errors can be eliminated by taking a sufficient number of repeated readings. When the quantity is being measured directly, these readings may simply be averaged. When the quantity of interest is not directly measured, but is computed from a formula involving two or more measured quantities, two methods of treating the data are available: (1) the data may be *faired*—i.e., the calculated values or functions thereof are plotted on graph paper as a function of any variable or variables occurring in the equation, and a smooth curve is drawn through the points; or (2) the more elaborate and rigorous *method of least squares* may be utilized to achieve the same result.[1-3]

Although *repetition* may be used as a method of reducing accidental errors, this is likely to be expensive and it is therefore important to consider to what extent duplicate readings will be useful.[1]

A single repetition commonly suffices for detection of blunders and accidents. The probability of an alert operator repeating the same mistake is usually small. Hence, if essentially the same reading is obtained the second time, it can reasonably be assumed that no major blunder has been made. The great importance of one repetition cannot be overemphasized. All readings should therefore be taken in *pairs*.[1]

In addition to exposing blunders, duplication also serves to reduce the probable accidental error to 70 per cent of its value for a single determination. As the number of repetitions is increased, however, they become rapidly less effective (see Fig. 3·1). Thus, two additional repetitions beyond the first pair reduce the probable accidental error only to 50 per cent of its value for a single determination. The average effectiveness for each reading in the second pair is only one-third of that for the first repetition. If the probable accidental error of a pair of tests is to be reduced 67 per cent by repetition, a total of 18 readings must be obtained, or 16 additional ones.[1]

When greater precision is required than can be secured from one pair of readings as planned, it is usually cheaper to achieve it by moderate improvements in technique. Multiple repetition is justified only:

(1), where the small increase in precision is necessary and better technique cannot be applied; and, (2), where repetition can be made very inexpensive.[1]

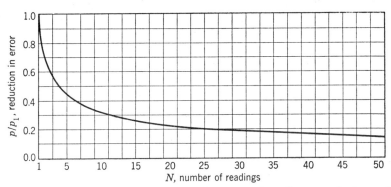

Fig. 3·1. Reduction of probable error with repeated readings. p_1 is the probable error of a single determination. p is the probable error of the mean for a series of repeated determinations. N is the number of equally careful independent determinations.

3·6 TREATMENT OF SYSTEMATIC ERRORS

Systematic errors, by definition, cannot be remedied by direct repetition with the same technique. In this case repetition is effective only when different technique is employed—i.e., different apparatus, method, operators, etc. When such a complete change is made, eliminating the procedure in which the systematic error is inherent, the same rules apply as for accidental errors.[1]

Thus, a single repetition usually suffices for detection of large errors. In the "heat-balance" method, the heat entering a system is compared to that which leaves, and any discrepancy is noted. Although even one repetition of this sort may nearly double the expense of the work, it is necessary if confidence is to be placed in the results.[1]

Reduction of probable systematic error by multiple repetition with successively changed apparatus is usually prohibitively expensive (see Fig. 3·1). When multiple apparatus is not used, the changing of one factor, for example, placing a number of thermocouples to read the same temperature and connecting them in series, would effectively constitute partial repetition, at small expense. Reduction of the probable systematic error by as much as 67 per cent through the use of additional thermocouples would, however, ordinarily be much too cumbersome. Where space permits, the use of one additional couple would be good technique.[1]

It should be emphasized, then, that one repetition by a different method is necessary as a means of detecting large systematic errors. Any further reduction in probable systematic error, however, should be achieved through improved technique.[1]

3·7 ACCUMULATION OF ERRORS

In relation to any one measurement, there are usually many contributing *sources* of error. If some of these individual errors are positive and some negative, they will tend to cancel one another, or they may be of like sign and tend to add in such a way that the total error is the numerical sum of the individual errors, p_1, p_2, p_3, etc. The probable error P is given by the formula

$$P = \sqrt{p_1^2 + p_2^2 + p_3^2 \cdots} \qquad (3·2)$$

This formula implies that the effects of errors diminish very rapidly with their size. An error one-third of another will add only one-ninth as much to the summation in Eq. 3·2. Errors only one-third of the larger ones may therefore be considered negligible, unless more numerous.[1]

While primary attention should be paid to the larger sources of error, there are usually a number N of these. If they are considered equal in size

$$P = p\sqrt{N} \qquad (3·3)$$

where p is the probable error permitted in each of the N individual sources.[1]

The number N will vary from one apparatus to another. However, N is usually larger than it is estimated to be, and the individual errors p are usually actually larger than they are assumed to be. In view of this and in order to provide a suitable factor of safety, it is considered good practice to limit individual errors to one-tenth the final allowable error; thus

$$a = 0.1A = 0.3P \qquad (3·4)$$

where a is the allowable magnitude of any single contributing source of error.[1]

REFERENCES

1. W. P. White, *The Modern Calorimeter*, pp. 26–36, 63, Chemical Catalog Co., New York, 1928.
2. G. C. Comstock, *An Elementary Treatise upon the Method of Least Squares*, pp. 12–16, Ginn and Co., New York, 1889.
3. A. G. Worthing and J. Geffner, *Treatment of Experimental Data*, 338 pp., John Wiley & Sons, New York, 1943.

4

CONDITIONS AFFECTING TEMPERATURE MEASUREMENT

4·1 THE PROBLEM

In any temperature-measurement undertaking, one should first ascertain the precision requirements (see Secs. 3·1 and 3·7), survey the general situation, and analyze the various factors involved in the *problem* at hand. This is preliminary to the quest for a solution to the problem and for the development of a satisfactory apparatus design.

A problem arises when it is desired to measure a given *temperature quantity* in a situation where a specific combination of conditions prevails. An example of one such condition is for the body in which it is desired to measure temperature to be in the *solid* state, as compared with its being in the *liquid* or in the *gaseous* state. The present volume of this work is limited to problems for bodies in the solid state. Bodies in the liquid and gaseous states, and in various special circumstances, will be dealt with in Vol. II. A description of other conditions that are pertinent to the solution of problems in *internal-temperature* measurement in solids will be included in this chapter.

The prevalence of such conditions is significant only in so far as these conditions affect the application of technique to measuring the temperature quantity of interest. Such a temperature quantity can be described by stating when, in time, and where, in a given body, temperature is to be measured, and with what degree of precision.

Thus, problems in the measurement of temperature are characterized by various combinations of conditions. Circumstances prevailing in a given situation, which constitute factors affecting the selection, design, or use of one or more of the component sections of the temperature-measurement technique, are termed *characteristic conditions*. The *solution* to a problem is a design that will suffice to measure the temperature quantity in the situation where this combination of conditions exists. Execution of this design will require that appropriate materials and techniques be available. Thus, if it is desired to measure temperature at an interior point in a body consisting of a very hard material, application of the design might involve a technique for drilling a hole in this material.

4·2 DESIGN COMPONENTS

Thermometric *design* is considered as consisting of three sections: (1), the *sensitive element;* (2), the *indicating instrumentation;* and (3), the communications or *leads.*

For example, with *thermocouples* the sensitive element is the "hot" *junction;* the indicating instruments include the potentiometer, standard cell, galvanometer, ice point, and switches; the leads are simply the actual lead wires running from the "hot" junction to the indicating instruments.

In *fluid-in-solid-container* thermometry the sensitive element is the bulb; the only indicating instrument is the pressure gage; and the leads consist of a tubular duct.

In *optical pyrometry* the one indicating instrument constitutes the entire design.

4·3 CHARACTERISTIC CONDITIONS

The remainder of this chapter will present a listing of examples of various categories of characteristic conditions. Succeeding chapters will discuss designs, materials, and techniques for coping with these conditions.

One of the classifications of characteristic conditions for temperature-measurement problems is the state of motion of the body, i.e., *static* or *moving.* A static or stationary solid body presents no special difficulties arising from this condition.

A moving body, for purposes of temperature-measurement design, is usually a body which is moving relatively to the indicating instruments. The motion of the body may be limited in range, as with periodic motion, or the motion may be continued in order to pass a given spot but once. The motions occurring in machinery are usually periodic.

In periodic motion the amplitude, frequency, shock accelerations, and internal vibrations are characteristic conditions. The amount of distortion imposed on the leads is determined by the amplitude. The amplitude and frequency together, plus the shock accelerations, determine the "whipping action" on the leads. The internal vibrations and shock accelerations determine the degree of mechanical ruggedness required in the sensitive element. Design must be such as to provide for operation under these conditions.

Another characteristic condition is *space*—that is to say, the dimensions and geometrical shapes of the body and its environment. For example, a shortage of space in the body itself may make it difficult to insert the sensitive element. A shortage of space in the vicinity of the body may cramp the leads.

The *nature of the medium* immediately surrounding the body is another characteristic condition. Thus, the leads may be required to pass through a jacket containing water, oil, or other fluid. The body may be inside a tank containing fluid under pressure, or the fluid may be electrically conducting or corrosive. It may be necessary that rigorously effective provisions be made to prevent leakage of the fluid at points where leads pass through the walls of such a jacket or tank. The body may be inside a furnace, requiring the leads to pass substantial distances through gases which may be at high temperatures, possibly also containing suspended particles of molten material. There also may be intense radiation. The space through which the leads must pass may be cold and damp so that moisture tends to condense on them. The body may be at a high electrical potential or the leads may be required to pass near objects at high potentials. Noise, building vibration, corrosive vapors in the air, cramped space, and similar factors may necessitate special kinds of indicating instruments or the locating of these at some distance from the sensitive element, thus necessitating the lengthening of the leads. Lengthening the leads might, as, for example, in thermocouple work, have the disadvantage of reducing the sensitivity.

The *temperature level* in the body is a characteristic condition. The sensitive element and portions of the leads must be adapted to withstand exposure to this temperature. The provisions necessary to achieve this adaptation may adversely affect precision.

The properties of the *materials* of which the body is composed are characteristic conditions. Machinability; fragility; electrical conductivity; susceptibility to peening, soldering, brazing, and welding may affect the ease with which the sensitive element can be installed. The thermal conductivity may affect the precision of the installation. The ease with which the material is corroded may affect the durability of the installation.

Temperature distribution and *variation* in the body, i.e., the variation of temperature from point to point in the body, are characteristic conditions. Thus, steep temperature gradients may necessitate increased exactness in the measurement of the location at which a given temperature is found to exist. Temperature changes with time may necessitate increased precision in the measurement of the instant at which a given temperature exists. The number of points at which simultaneous measurements are required will affect the difficulty of achieving this precision.

The precision requirements are a characteristic condition and affect the choice of instrumentation throughout.

5

THE THERMOCOUPLE THERMOMETER—CIRCUITS

The thermocouple technique is the most widely useful method of measuring internal temperatures in solid bodies. It is intended here to describe fully the techniques that are most suitable for various purposes with the precautions which must be taken to obtain results of definite degrees of precision. It will be obvious that, for much work, many of these precautions may be omitted, or relaxed.

5·1 THERMOELECTRIC POWER

A simple thermocouple circuit is shown in Fig. 5·1. Here, two metals, A and B, form an electric circuit with *junctions* that have temperatures t_1 and t_2. In general, if the junction temperatures t_1 and t_2 are not identical, an electromotive force, or emf, E, will exist in such a circuit. The magnitude of the emf will depend on the metals used, on the amount of the temperature difference, $t_1 - t_2$, and on the values of the actual temperatures, t_1 and t_2. By including a suitable device to indicate any electromotive force or flow of current that may occur in the circuit, the temperature difference $t_1 - t_2$ can be measured.

The term *thermoelectric power*, e, as applied to such a circuit, is defined, for a given pair of metals and a specified average temperature, as the ratio of the magnitude of the thermoelectric emf, E, to the temperature difference, $t_1 - t_2$, between the junctions.

If the thermoelectric power for a pair of metals A and B at junction temperatures t_1 and t_2 is e_{AB}, and the thermoelectric

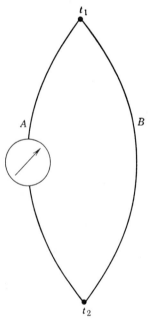

Fig. 5·1. Thermocouple circuit.

power for the pair A and C at the same pair of temperatures is e_{AC}, then the thermoelectric power at these temperatures for the pair B and C is

$$e_{BC} = e_{AC} - e_{AB} \qquad (5 \cdot 1)$$

Here B replaces A in Fig. 5·1 and C replaces B. The signs of e_{AB}, e_{AC}, and e_{BC} are taken as positive if the emf is in such a direction as to drive the current in a clockwise direction in Fig. 5·1. We may think of A as a standard or *reference* metal. Then, if t_2 is the lower temperature, we have the convention that the sign of the thermoelectric power is taken as positive if the current tends to flow from the given metal to the reference metal at the cold junction. In the published tables of thermoelectric powers for various metals, lead is usually taken as the reference metal.

For various pairs of materials commonly used for thermocouples, such as iron against constantan, chromel P against alumel, platinum against platinum-rhodium alloy, and copper against constantan, very complete tables are published. These assume a fixed temperature for the cold junction t_2, i.e., 0°C or 32°F. The thermoelectric emf in volts is given for each value of t_1 in small steps and over a wide range. The thermoelectric emf can be calculated for other temperatures of the cold junction besides those given in such published tables. Thus, suppose the emf given for $t_2 = 32°F$ and $t_1 = 212°F$ is E_{32-212}, and the emf for $t_2 = 32°F$ and $t_1 = 213°F$ is E_{32-213}, then the emf for $t_2 = 212°F$ and $t_1 = 213°F$ is

$$E_{212-213} = E_{32-213} - E_{32-212} = e \qquad (5 \cdot 2)$$

Thus, also, the thermoelectric power at any temperature is simply the tabular difference for 1°F at that temperature.[1]

5·2 MULTIPLE-METAL CIRCUITS

A thermocouple circuit may contain four metals, as shown in Fig. 5·2. This is the case when the leads are, for practical reasons, made of materials different from those used for the thermocouple proper.[2] Let E_{ABDC} be the thermoelectric emf for this circuit. Also, let E_{AB} be the thermoelectric emf for metals A and B for hot-junction temperature t_1 and cold-junction temperature t_i. Similarly, let E_{CD} be that for metals C and D at hot-junction temperature t_i and cold-junction temperature t_2. Then

$$E_{ABDC} = E_{AB} + E_{CD} \qquad (5 \cdot 3)$$

Let e_{AB} be the average thermoelectric power over the temperature range, t_1 to t_i, for metals, A and B, and e_{CD} that for metals, C and D, over the range, t_i to t_2. Then $E_{AB} = e_{AB} (t_1 - t_i)$, and $E_{CD} = e_{CD} (t_i - t_2)$. Also

$$E_{ABDC} = e_{AB}(t_1 - t_2) - (e_{AB} - e_{CD})(t_i - t_2) \qquad (5 \cdot 4)$$

If metals C and D are alike, as with copper leads, then $E_{CD} = 0$, $E_{ABDC} = E_{AB}$, and t_i is effectively the cold junction. If t_i and t_2 are the same, E_{CD} is likewise zero, and $E_{ABDC} = E_{AB}$.

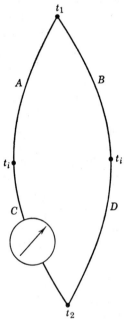 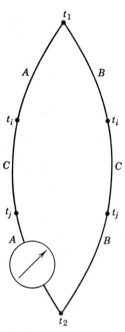

Fig. 5·2. Thermocouple circuit with four metals.

Fig. 5·3. Thermocouple circuit with three metals.

The first term in Eq. 5·4 gives the thermoelectric emf that would result if the circuit consisted exclusively of metals A and B. The second term may be negligible if the thermoelectric powers, e_{AB} and e_{CD}, or the temperatures, t_i and t_2, are sufficiently alike. This fact makes it possible, in certain types of work, to use cheaper or more rugged materials for the lead wires. The second term usually represents an error. The magnitude of this error can be estimated by noting the effect of arbitrarily varying t_i.

Suppose that a thermocouple circuit were built containing three metals as indicated in Fig. 5·3. Let e_{AB} be the average thermoelectric power over the temperature ranges, t_1 to t_i, and t_j to t_2. Then

$$E_{ACABCB} = e_{AB}(t_1 - t_2) - e_{AB}(t_i - t_j) \qquad (5 \cdot 5)$$

Thus, if a portion, $(t_i - t_j)$, of the temperature difference, $t_1 - t_2$, occurs in a section in which both leads are of the same material, the readings will be lower by the amount, $t_i - t_j$.[1]

5·3 PARASITIC THERMOELECTRIC EMFS

In thermoelectric circuits the temperature always varies along the wires from the cold to the hot junction. The *law of intermediate temperatures* states: *if the materials, A, B, C, D, etc., in Figs. 5·1, 2, etc., are uniform throughout in their thermoelectric properties, the thermoelectric emfs are dependent solely on the junction temperatures,* t_1, t_2, t_i, *etc.* Such uniformity, or *homogeneity*, was assumed in Eqs. 5·1, 2, 3, 4, 5.

Equation 5·4 shows the effect of a change in properties at one point having a temperature t_i. Thus here, if $e_{AB} - e_{CD}$ is 15 per cent of e_{AB} and if $t_i - t_2$ is one-fourth of the total temperature difference, $t_1 - t_2$, a 3.75 per cent error is introduced. If $t_1 - t_2$ is 1000°F, this error would be 37.5°F and would drift in magnitude with any random changes in t_i.

If there are many such changes in properties, even though individually small, at a corresponding variety of temperatures, the resulting effect is complex. The smaller the variations in temperature, the smaller will be the resulting error for a given degree of *inhomogeneity* in the metals. It is customary to use wires that are produced or tested to have sufficient homogeneity to meet the requirements of the given situation. All joints should be fused to avoid inhomogeneity which would result from the use of any solder or brazing material.

Thermoelectric circuits always include *indicating instruments*, switches, etc. The portion of the circuit through any such device nearly always consists of several metals. The *law of intermediate metals* states: *if in any circuit of solid conductors the temperature is uniform from any point P through all the conducting matter to a point Q, the thermoelectric emf in the circuit is the same as if P and Q were put in contact.*

Thus, if the temperature throughout any such switch or indicating device is uniform, no error will result. Likewise, the smaller the variations in the thermoelectric properties among the various metals used, the smaller the resulting error will be for a given degree of nonuniformity in temperature within the group of switches and indicating devices. The metals for this portion of the circuit are customarily limited to copper, brass, lead-tin solder, silver, and manganin, since the thermoelectric properties of these metals do not differ greatly, i.e., they have low thermoelectric powers with respect to one another. The temperature within this portion of the circuit is then maintained at a degree of uniformity corresponding to the required precision.

It is usually sufficient for the switches and instruments to be kept in a well-constructed room provided with ordinary thermostating (i.e., $\pm 4°F$), and at a distance of 6 feet or more from any hot or cold surface such as a window, radiator, or poorly insulated steam pipe, and not exposed to any drafts or direct sunshine. In very precise work (i.e., to within ± 0.01 or $\pm 0.001°F$) additional precautions may be required, such as packing the instruments in cotton or immersing them in constant-temperature oil baths. If instruments must be placed in thermally exposed positions, protection by some adequate form of thermal insulation is necessary even in low-precision work. The uncertainty in measurement, because of inadequate protection of the instruments, can be estimated by noting the reading when the thermocouple is shorted out.[1]

5·4 PARASITIC VOLTAIC EMFS

Besides the types of error that result from variations in temperature and composition in a thermocouple and are usually termed *parasitic* or *spurious thermals*, errors can also result from what are called *parasitic* or *spurious voltaic effects*. Whenever two metals of different composition are in electrical communication through some form of *electrolyte*, a *voltaic cell* is established. This principle is applied in the ordinary dry cell used for flashlights, etc. The emfs of such cells are of the order of magnitude of 1 volt. Thermoelectric emfs are measured in millivolts or microvolts, i.e., thousandths and millionths of volts, respectively. For iron and constantan wires, about 30 microvolts represent 1°F. Thus, any incipient voltaic effect can readily result in an error of magnitude sufficient to be serious even in low-precision work; whereas errors large enough entirely to mask the thermoelectric voltage are possible.

Voltaic effects can occur in thermocouple circuits where there is an insulation leak in the presence of moisture. Such a leak often occurs through semiporous insulation in which moisture has condensed or over the surface of an impervious insulator on which moisture has condensed. Although pure water is a moderately good electrical insulator, such moisture usually contains dissolved materials which render it semiconducting. If traces of acid or acid-forming salts are present, as might result from using acid flux in soldering, the conductivity of any condensed moisture is greatly increased. Such voltaic effects can occur even when the difference in composition between the two metals is very small, as, for example, when the leak is between two points only slightly distant from one another on the same wire.

Since voltaic effects usually result from electrical leakage through moisture, the most important precaution is to keep the circuit dry. The high-temperature portions of the circuit usually cause no trouble in this respect. Trouble is most likely to arise in those parts which are below room temperature, as, for example, in the vicinity of the ice point. At high values of relative humidity, appreciable condensation can occur on objects at room temperature or even somewhat above room temperature.

The precautions to be taken against voltaic effects (and such precautions are always to be regarded as essential) are as follows: (1) perform all soldering with rosin flux and avoid exposing any parts of the circuit to acids or alkalies for any reason whatever, even though the parts are thoroughly washed afterwards; (2) make all reasonable or necessary efforts to keep the room dry—i.e., maintain low relative humidity; (3) guard against possible porosity in insulation in parts of the circuit not at high temperatures by thorough coating with some impervious material. Various excellent varnishes are available. Where extreme precision is required and in other cases where it is convenient to do so, the wires are often embedded in paraffin; and (4) surface conductivity over the insulating members at switches where portions of the conductor are exposed (unless the entire switch is immersed in oil) is best coped with by using, as insulators, impervious materials that normally have low surface conductivities, and by keeping the insulator surfaces clean. This precaution tends to cause any moisture which condenses to have less conductivity. Natural ceresin wax is a material of unusually low surface conductivity and may be used successfully for insulation in precision work even where the relative humidity is high. It remains rigid at high room temperatures and is fairly satisfactory mechanically. Various moisture-repellent coatings are available.

5·5 THERMOCOUPLE WIRE

Four pairs of thermocouple wire materials are widely used. These are copper against constantan * (60 per cent copper with 40 per cent

---

* The name, constantan, is also used to refer to other copper-nickel thermocouple alloys of varying compositions (thus containing up to approximately 45 per cent nickel with 55 per cent copper), although these have separate trade names.

nickel), iron against constantan, chromel P (90 per cent nickel with 10 per cent chromium) against alumel (95 per cent nickel with aluminum, silicon, and manganese comprising the remainder), and platinum against platinum-rhodium (90 per cent platinum with 10 per cent rhodium, or 87 per cent platinum with 13 per cent rhodium).

Standard tables for thermoelectric emfs have been established for these four combinations of materials.* Matched pairs of wires are furnished commercially to conform, within manufacturing tolerances, to these tables. Manufacturers of thermocouple wire also furnish tables applying specifically to their own products.[3-6]

Iron against constantan, in heavy gages, i.e., No. 8 B & S gage (0.128 in.) or larger, and suitably protected from corrosion, can be used at temperatures as high as 1500 or 1800°F, depending on the requirements as to precision and lifetime in service. Bare, unprotected wires in the smaller sizes, i.e., No. 24 to 30 B & S gage (0.020 to 0.010 in.), should not be used above 600 or 1050°F. The lifetime in service at a given precision will be greater at 600 than at 1050°F. Rapid deterioration begins at 1050°F. In the absence of protection, either oxidizing or reducing atmospheres, preferably reducing, are used. It is feasible to use iron against constantan at temperatures as low as −423°F. The two materials are both mechanically strong and tough and of moderately low thermal conductivity (see Sec. 7·6). The thermoelectric power is high and nearly uniform over a wide temperature range, i.e., 32.2 μv (or microvolts) per °F average over the usual working range, 32 to 1650°F, or 30.0 μv per °F in the range, 32 to 212°F. Matched wires in impregnated, glass-fiber-braid duplex cables are commercially available.[1, 3, 6-8]

Copper against constantan is widely used at temperatures below 32°F, and is suitable for application at temperatures as low as −423°F. In heavy gages, i.e., No. 14 B & S gage (0.064 in.) or larger, and protected from corrosion, this combination can be used at temperatures as high as 700 or 1100°F, or, bare in the smaller sizes, i.e., Nos. 20 to 30 B & S gage (0.032 to 0.010 in.), up to 350 or 600°F, depending on the requirements as to precision and lifetime in service. The copper strand is subject to the disadvantage of high thermal conductivity. The ease with which copper can be fabricated in a pure, homogeneous condition tends toward reliability in this couple. In the absence of protection, either oxidizing or reducing atmospheres are used. The thermoelectric

* At the time this goes to press, standardization has not yet been accepted for iron against constantan. Several tables continue in use.

Sec. 5·5　THERMOCOUPLE WIRE　　41

power is high, i.e., 23.8 μv per °F in the range 32 to 212°F and 9.8 μv per °F at −300°F.[1,4,7,9,10]

Chromel P against alumel, in heavy gages, i.e., No. 8 B & S (0.128 in.) or larger and suitably protected from corrosion, can be used at temperatures from −300°F to as high as 2200 or 2450°F; whereas the smaller sizes, i.e., No. 30 B & S gage (0.010 in.) or larger, can be used bare at 1200°F. In the range 1000 to 1800°F, this combination possesses a somewhat longer lifetime in service than iron against constantan. The two materials are both of low thermal conductivity and when new they are strong and tough. However, contamination in service results in brittleness which is more severe in the alumel than in the chromel P. In the absence of protection, oxidizing atmospheres must be used. The thermoelectric power is high and nearly uniform over a wide temperature range, i.e., 22.4 μv per °F average over the usual working range, 900 to 2200°F, or 22.8 μv per °F in the range 32 to 212°F.[1,4,8,11,12]

The platinum against platinum with 10 per cent rhodium thermocouple is specified in the International Temperature Scale for the temperature range, 1166.9 to 1945.4°F. Despite the reliability and reproducibility of these materials, this couple is not usually used industrially in this temperature range because of its relatively low thermoelectric power. Since the thermoelectric power passes through a zero value at −216.4°F and is only about 2.8 μv per °F at 32°F, this combination is never used below 32°F. It can, however, be used at temperatures as high as 2700 or 3100°F, depending on the requirements as to precision and lifetime in service. The thermoelectric power at these higher temperatures is still low, i.e., 6.6 μv per °F average over the working range 2000 to 2900°F. The combination platinum against platinum with 13 per cent rhodium has a slightly higher thermoelectric power, i.e., 7.7 μv per °F in the working range 2000 to 2900°F. Clean air serves as a protective atmosphere. Exposure, at even moderately elevated temperatures, to oxidizable materials, such as carbon, hydrogen, sulfur, phosphorus, and their oxidizable compounds; to metallic vapors in reducing atmospheres; and to compounds containing silicon, such as porcelain, mica, and soapstone, must be avoided. Brittleness, sufficiently severe to cause failure, may result from serious contamination; whereas errors up to 5 per cent, i.e., 125°F at 2500°F may occur prior to such failure. Thus protection tubes are used. Both materials are ductile but become exceedingly soft and weak at 3100°F, requiring continuous support. The thermal conductivities of both materials are moderately low.[1,4,7,13–15]

Another combination used is constantan against silver. The thermoelectric power does not differ greatly from that of constantan against copper. However, operation at temperatures up to 1100°F is feasible. The silver strand has the disadvantage of high thermal conductivity.[7]

Standard tables are given for chromel P against constantan. The thermoelectric power is high, i.e., 35.1 μv per °F average in the range, 32 to 212°F; whereas the greater tendency to brittleness in the alumel member of the chromel P against alumel couple is avoided. The highest temperatures are limited by the constantan to somewhat above those feasible for the iron-against-constantan combination.[4,12]

Bismuth against bismuth alloy "B" has very high thermoelectric power, i.e., 40.9 μv per °F in the range −114 to 350°F, which is also uniform over the working range −424 to 350°F. Both thermal capacities and conductivities are very low. Reasonably ductile wires are available, but care is required in use.[16]

For temperatures above 3100°F, less satisfactory materials are available. Thus, iridium against iridium with 10 per cent of ruthenium can be used up to 3600°F. This couple is, however, brittle and deteriorates rapidly because of oxidation of the ruthenium. Calibration data are available, the thermoelectric power being very low, i.e., 0.9 μv per °F average in the range 2900 to 3250°F. Substitution of iridium with 10 per cent rhodium for the iridium member increases the thermoelectric power, while still permitting operation up to 3600°F.[7,17,18]

Iridium against 60 per cent rhodium with 40 per cent iridium can also be used up to 3600°F and under oxidizing conditions. These materials are less brittle. Calibration data are available, the thermoelectric power being low, i.e., 2.3 μv per °F average in the working range 2900°F to 3600°F.[7,17,18]

Tungsten against iridium can be used up to 3800°F. A neutral protective atmosphere, or operation in vacuo, is required. Because of the use of pure materials deterioration is less rapid. Calibration data are available, the thermoelectric power being relatively high and uniform over the working range 2900 to 3800°F, i.e., 14.3 μv per °F.[18]

Tungsten against tungsten with 25 per cent molybdenum can be used up to 5400°F. Strong, ductile wires are available, although tungsten tends to become brittle. A neutral protective atmosphere, or operation in vacuo, is required. Calibration curves are available, the thermoelectric power being low, i.e., 1.9 μv per °F average in the working range 3800 to 5400°F.[7,17-21]

Carbon against silicon carbide [22] and carbon against boron carbide [23] have been proposed for high-temperature use.

Sec. 5·6 TESTING THERMOCOUPLE WIRE

At temperatures below $-423°F$ the thermoelectric powers of widely used combinations, such as copper against constantan, become very small, but less familiar materials are available. Thus, gold with 0.30 per cent of cobalt against silver with 1.8 per cent of gold can be used at temperatures as low as that of liquid helium and has a thermoelectric power of 8.35 μv per °F at $-427°F$ (32.7°R). Similarly, silver with 0.38 per cent of gold against copper with 0.088 per cent of iron has a thermoelectric power of 8.82 μv per °F at $-427°F$ and 2.97 μv per °F at $-456°F$ (3.7°R). Calibration data are available.[9,16,17]

5·6 TESTING THERMOCOUPLE WIRE

Wire furnished by reputable firms for use in thermocouples is now being produced sufficiently uniform for most purposes. It is therefore ordinarily not necessary to check new wire for uniformity.[24] Where wire has been exposed to corrosive conditions, such as to gases at elevated temperatures, where it has been necessary to splice strands of wire coming from different spools, or where high precision is required, a check should be made.[8,12-15]

It is impracticable or impossible to make any sort of measurement of an inhomogeneous condition in thermocouple wire such as to permit computation of corrections for this effect. Methods are available only for the detection of inhomogeneity and to estimate a resulting degree of *uncertainty* in temperature measurement.[24]

To detect abrupt changes in thermoelectric power, connect the two ends of the wire to terminals of a sensitive galvanometer. Heat locally, establishing steep temperature gradients in the wire at the splice or other portions of the wire suspected. Note how much the galvanometer deflects when the heat is applied at the various points. It is often possible to perform this test on the installed circuit, using the regular indicating instrumentation.[24]

To estimate the degree of uncertainty in temperature measurement to be expected from inhomogeneities detected by this test, it is necessary to know the voltage sensitivity of the galvanometer used and the local temperature increase applied in heating the wire. Then

$$\Delta t = E t_2 / e t_1 \qquad (5 \cdot 6)$$

where Δt is the uncertainty in temperature measurement due to inhomogeneity, °F; E is the emf corresponding to the largest galvanometer deflection found in this test, μv; e is the thermoelectric power of the thermocouple used, μv per °F; t_1 is the temperature rise produced in local heating, °F; and t_2 is the largest temperature difference occurring between any two points in the thermocouple circuit in operation, °F.[24]

Δt is not necessarily the largest error that could result from inhomogeneity in such a case. However, errors as much as several times Δt would probably be unusual, whereas the actual error might be substantially less than Δt.

To detect gradual changes in thermoelectric power, this method should be modified. Instead, the wire is doubled and the loop inserted to various depths in a uniformly heated furnace, or flask of liquid air. The emf, E, then results from the difference in thermoelectric properties in the wire at the two points of entry into the furnace. t_1 is the temperature excess of the furnace above, or of the liquid air below, room temperature, and Δt is computed by Eq. 5·6.[24]

After a reasonable degree of homogeneity in one sample of wire has been established, this sample can be used as a standard in testing the homogeneity of similar wires by welding the two together and inserting the junction into a uniformly heated furnace, or flask of liquid air. The emf, E, then results from the deviation in the thermoelectric properties of the test wire at the point of entry into the furnace. Δt is computed from Eq. 5·6.[24]

Any wire found to have inhomogeneities such that Δt is more than one-tenth of the allowable error (see Secs. 3·3, 7) should be discarded.

5·7 CALIBRATION OF THERMOCOUPLE WIRE

For measuring temperatures in the range -300 to $3100°F$, thermocouple wire is available commercially in matched pairs such as to conform, within specified tolerances, to the published standard tables. Each strand of wire, as produced, is calibrated. Selected strands of the two materials are then paired such that the temperature-emf relationship for each such pair does not deviate by more than the stated amount. Common tolerances are ± 0.25 per cent to ± 0.75 per cent, although manufacturers usually also refuse to assume responsibility for absolute errors smaller than some minimum amount, such as 0.1 to 15°F.[3, 6, 11, 25]

These precisions are sufficient for a wide range of technical work. If greater precision is needed, several alternatives are available. Thus, manufacturers will furnish, at an additional charge, paired wire calibrated at temperatures specified by the user. Deviations from the standard tables at these temperatures are given as correction data marked on tags attached to the spools of the wire.[6]

Manufacturers will also furnish, at an additional charge, selected paired wire conforming with the standard tables to within much closer than the usual tolerances. Such wire is also tagged with correction data.[6]

In case an independent calibration of maximum authenticity and precision is required, the National Bureau of Standards (Washington, D. C.) regularly furnishes the necessary service at a charge sufficient to compensate for costs. Temperature ranges from -321 to $2850°F$ are covered. A minimum length of wire, such as 24 or 36 in., must be supplied. The charge depends on the number of temperatures checked, on whether they are checked against a standard thermocouple or against fixed points, and on the temperature range in which the couple is checked. A certificate is furnished stating the emf at each of the temperatures checked. The calibration will not be undertaken if the thermocouple proves to be such that it will not yield the specified accuracy. In such cases, a report is issued giving the results obtained. A fee, corresponding to the costs of the work performed, is charged.[26]

Laboratory calibrations of thermocouples made from inhomogeneous wire are meaningless. The emf developed by such thermocouples depends, not only on the junction temperatures, but also on the temperature patterns along the length of the wires. Such temperature patterns will, in general, not be the same in service as in the calibration rig. In consequence, the temperature-emf relationship in service will not be the same as that found in the calibration. If, in service, the temperature pattern is always the same at a given temperature level, a calibration may be performed installed, if it is also possible to insert a certified thermometer to measure the same actual temperature.[24, 27]

Laboratory calibration has only to do with wire that is found to be homogeneous in thermoelectric properties along its length. As such it is not necessary to calibrate every couple, individually, which may be made up out of a given pair of matched strands of wire. Thus couples can be taken from each end of a pair of strands, or, if the length is great, at suitable intervals. Any discrepancies among the calibration data for these thermocouples made from the same pair of strands will be a measure of corresponding gradual change in thermoelectric properties along the length, i.e., of that variety of inhomogeneity. If these discrepancies amount to more than one-tenth the allowable error, the entire spool should be discarded.

Thus, the cost of certification is spread over an entire spool and does not apply to each individual couple. It is furthermore inconvenient to calibrate individual couples. Different lengths of leads may be required in the service installation than are suitable in the laboratory calibration work. This might necessitate subsequent *splicing*. It may be necessary to remove the insulation for calibration.

Frequently, the interest lies more in the measurement of temperature differences, or in the determination of a temperature pattern, than in learning the absolute values of these temperatures on the International Temperature Scale. In this case, all thermocouples for the job should be made from the same pair of strands. Then the error in measuring temperature difference will merely be that in the thermoelectric power applied to the given temperature difference.

For the measurement of temperatures below −321°F and above 3100°F matched, calibrated thermocouple wires are not at present commercially available.[6] Similarly, in these temperature ranges, certification service is not regularly rendered by the National Bureau of Standards (Washington, D. C.).[26] Wire from suitable pure materials and alloys may require its being drawn specially or its being purchased from stocks not intended for thermoelectric application. Such wire will usually require testing for homogeneity and calibration by the user.

Above 3100°F, such wire can be calibrated against the International Temperature Scale by immersing the "hot" junction and leads to a sufficient length in a uniform-temperature enclosure. The temperature of this enclosure can then be measured with a monochromatic optical pyrometer. Fixed points are also available.

Provision of the enclosure for the temperature range 3100 to 4500°F requires special apparatus. Tungsten heater wire, self-supported in its heated length and operated in an atmosphere of pure argon, can be used to heat the interior of an insulated furnace wall lined with slabs of such highly refractory material as fused zirconia.[28] A cooled window is required for sighting with the optical pyrometer.[18]

For temperatures above 4500°F, the effective enclosure can be the heater filament itself. This can be of tungsten ribbon stock staggered so as to form an "optical wall" or *radiation shield* parallel to and around the thermocouple wires. A hole or slit in this radiation shield will permit sighting on the thermocouple junction with the optical pyrometer. The outside envelope of this device is made of metal and water-jacketed. Pure argon is used as a protective atmosphere. This should not be in too rapid circulation to avoid excessive convective cooling of the thermocouple, which is then heated by radiation nearly to thermal equilibrium with the inner surface of the enclosure. Such an enclosure can be operated up to 5400°F.[18, 21]

Below −321°F, thermocouple wire can be calibrated against the Provisional Temperature Scale (see Sec. 1·15) by reference to fixed points in a cryostat. Suitable cryostats are available commercially.[29]

Calibration of thermocouples in the range −321 to 2850°F is most conveniently performed by welding the test junction into the same bead with that of a certified couple. When this composite junction is held at a reasonably steady temperature, simultaneous readings on the two couples effect the comparison. A large assortment of suitable furnaces, thermostated baths, and cryostats is available commercially.[24, 29]

The ultimate in precision is, of course, only attained by calibration and frequent recalibration of the individual couple against primary fixed points. The same temperature pattern must prevail along the wires as that occurring in service. All parasitic emfs must have been reduced to a minimum and their residual magnitudes rendered nearly the same as in service. All indicating instrumentation must have been calibrated with similar care. For this technique the reader is referred to "Methods of Testing Thermocouples and Thermocouple Materials" by Roeser and Wensel, with attention to their bibliography.[24]

The ultimate in precision is usually not, however, necessary. The precision actually required should be decided upon in planning a project (see Sec. 3·1), and adhered to thereafter.

5·8 JUNCTIONS AND SPLICES

The term "junction" applies to each of the electrically conducting joints made between the thermocouple wires to form the thermocouple. The term "splice" is used when the conducting joint is made between two wires of the same, or thermoelectrically equivalent, metals.

A junction can be made in any of several ways: (1), the wires can be fused together in a beaded or butt-welded junction; (2), they can be brazed, silver-soldered, or soft-soldered together; (3), the two wires can be dipped into a pool of mercury or other molten metal; (4), both wires can be soldered, brazed, or fused to solid metal; and, (5), both wires can be clamped under a *binding* post.

In fused junctions there may be a layer of intersolution separating the two metals. In the brazed and soldered junctions there is a separating layer of the brazing or soldering material. The wires in soldered or brazed junctions may also be coated with the brazing or soldering material for a distance back from the junction. Where the wires dip into or are joined to a metallic conducting medium, a section of this medium separates the two metals. The process of forming the junction may thermoelectrically alter the metals near the junction by corrosion or contamination, or by changing the crystal structure. If they are dipped into a molten metal, gradual contamination by intersolution may result.

Thus, in general, there is a portion of the circuit, in the neighborhood of the junction, that consists of material other than either of the original thermocouple metals. This can be regarded as a short length of wire composed of a third metal, or as an inhomogeneous section. If this portion of the circuit is at the same, uniform temperature, the effect will be the same as though the two metals were joined directly at the point of entry into the uniform temperature region. This point will then be the effective junction. If the temperature within this inhomogeneous portion of the circuit at the junction is not uniform, but varies over a certain range, it will be uncertain as to what temperature within (or possibly beyond) that range will be indicated by the junction. Thus, the effective location of the junction in the circuit will be uncertain. To minimize such uncertainty when the temperature is nonuniform in the vicinity of the junction, it is necessary that the inhomogeneous portion at the junction be made negligible in size or at least as small as feasible. This is best achieved in the beaded or butt-welded junction. Brazed or soldered junctions are permissible when the temperature is uniform in the vicinity of the junction.

Where a splice is made between two wires of the same metal or of metals intended to be thermoelectrically equivalent, as in the lead wires, in a region where the temperature varies along their lengths, the splice should be made by fusing the two ends together in a butt weld or fused splice. This avoids parasitic thermoelectric effects which might result from any intermediate layer of solder or brazing material. Clamping the two wires directly together is also satisfactory if good contact is made.[30]

Figure 5·4 shows a brazed or soldered junction. To make such a junction, the ends of the two wires are bared and cleaned with fine emery cloth. They are then twisted together, as shown, and cleaned by dipping in carbon tetrachloride. If lead-tin solder is to be used, the twisted portion is pressed into a block of rosin by a hot soldering iron, *tinned* with a just sufficient amount of the molten solder. Slight rubbing of the wires with the iron will result in an adequate flow of solder to *wet* the wires. If silver solder is used, the twisted portion is dipped into flux No. 43, manufactured by Krembs and Company (Chicago, Illinois), or into any other suitable flux,[31] which is mixed with water to a creamy consistency. A flame is then applied to the flux until it dries, melts, and flows over the surface. When at a dull red heat, the end of a strip of silver solder wire is brought in contact

Fig. 5·4. Brazed or soldered junction.

just long enough to fill the interstices. This flux can be washed off with hot water. The flame can be one of air or oxygen with coal gas, natural gas, or hydrogen. Brazing can be performed similarly to silver-soldering.[32]

Gas flames applied to platinum, platinum alloys, and alumel tend to produce contamination, with resultant inhomogeneity and brittleness. Electric-fusing is best adapted to these materials.[8, 12-15]

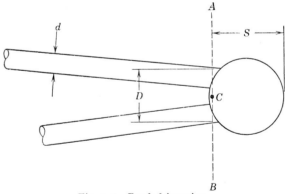

Fig. 5·5. Beaded junction.

Figure 5·5 shows a beaded junction. The effective location of the junction in the circuit is in the plane AB. If the temperature of the metal is not uniform in plane AB, the region of uncertainty is D in length and d in width. D can be made about $2\frac{1}{2}$ times the wire diameter d. The effective junction is usually assumed to be at the midpoint C.[33, 34]

Beaded junctions can be made by the arc method, by the condenser-discharge method, or by the gas-welding method. The technique for the *arc method* is as follows.

1. Grip the two strands of wire in smooth-jaw pliers so that they lie parallel and close to each other and with a free projection of about 15 wire diameters.

2. With a second pair of pliers make a tight twist of the projecting ends. It is somewhat difficult to make a successful hand twist of very

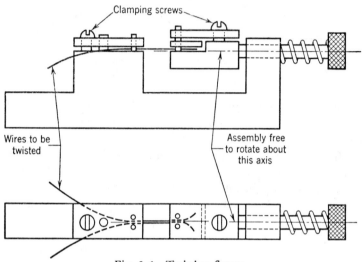

Fig. 5·6. Twisting fixture.

fine wires. A special device, shown in Fig. 5·6, is used for wires smaller than No. 30 B & S gage (0.010 in.).

3. Using sharp scissors, clip off all but one and one-half to two turns of the twist.

4. Weld the projecting twist into a spherical bead. Use d-c current with amperage as determined by previous trial. Use an adjustable resistance in the circuit. When the best setting has been found, note the value of the short-circuit current for future work. Thus, for example, good beads are formed on 0.010-in. wires when the resistance is adjusted to give a short-circuit current of about 4 amp.

Grip the wires in smooth-jaw pliers or in a special vise as close to the twist as possible, and straighten the projecting twisted portion. Make the circuit connections such that the wires are positive, and a carbon or graphite rod negative. Set the rheostat to the predetermined value and bring the carbon rod up to the end of the twist so as to touch it lightly and momentarily.

5. Examine the bead formed under a magnifying glass. If the fusing is insufficient, the arcing operation should be repeated.

With a little practice smooth, spherical beads of the required diameter can be made, as shown in Fig. 5·5.

A mercury pool covered with lubricating oil can be used in place of the carbon rod. In this case, however, because of the obscuring effect

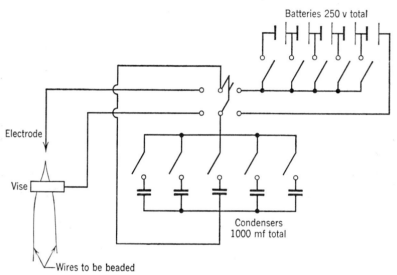

Fig. 5·7. Circuit for beading by condenser-discharge method.

of the oil, it is more difficult to see what one is doing, and more skill is required to produce satisfactory beads. The oil is intended to protect the heated portion adjacent to the bead. However, with alumel and the platinum alloys, it will be expected to result in contamination and brittleness.

Although the arc method is usually preferable, beading can also be performed by the *condenser-discharge method*. Here, the bare portion of the wire is gripped in a pin vise which is electrically connected through a heavy copper lead to one pole of a bank of condensers (see Fig. 5·7). A movable copper electrode is similarly wired to the other pole. The condensers are charged to a suitable voltage with heavy-duty-type radio batteries. After being disconnected from the batteries by opening a switch, the condensers are discharged by drawing a spark between the twisted ends of the wires and the copper electrode. If, after examination, it is found that not all the twisted portion has been absorbed by the bead, one or more additional discharges can be applied. For 0.010-in. diameter iron and constantan wires, 1000-μf (microfarads) capacity and 125-v (volts) d-c are used.

Since the heat released for welding is predetermined in quantity by the voltage and capacitance used, the condenser-discharge method requires a minimum of skill. It is therefore valuable in beading very fine wires. The weld produced is, however, more fragile than that made in an arc and is inclined to be more irregular in shape.

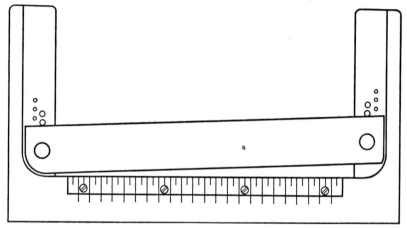

Fig. 5·8. Junction gage.

In the *gas-welding method* a small oxygen-illuminating gas flame is used without flux to fuse the ends of the wire together into a bead. This method should not be used on platinum or platinum alloys (see above).

After the bead has been made, the distance s, in Fig. 5·5, from the tip of the bead to the plane AB, can be measured with the junction gage shown in Fig. 5·8. The junction is mounted beneath a sheet of waxed paper in order to be illuminated from below. The junction gage is then slid about on the paper until the shadow of the bead in the direction of s (see Fig. 5·5) is just included between the inclined straightedges. The distance, s, is then read directly from the scale. Readings can be made accurately to within ±0.001 in.

A pair of butt-welded junctions is shown in Fig. 5·9. Such welds can be made in wires as small as No. 40 B & S gage (0.003 in.) and in all larger sizes. The interface between the two metals is sharply defined and usually almost perpendicular to the wire. The welds are no larger than the original wire and nearly as strong. Equipment for the making of such welds is available from the Micro Products Company (Chicago, Illinois). However, a small jig to align the wires and to bring the ends into contact at a definite and reproducible pressure, with a suitable commercial spot welder equipped with a timer, is all that is required. A shielding jet of argon or helium can be used to protect the wire during welding.[35] Before butt-welding, the ends of the two wires are bared for a short distance and squared with a fine file. After welding in the jig, the weld can be cleaned by rubbing with fine emery paper.

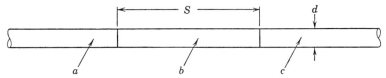

Fig. 5·9. Butt-welded junction. a, metal A; b, metal B; c, metal A.

It may be desired to make a difference or gradient couple which involves production of a second weld at a specified distance s from the first (see Fig. 5·9). To make the second weld, proceed as follows.

1. For wires of iron and constantan, paint the weld with a thin mix of Technical "B" Copper Cement, produced by the W. V-B. Ames Company (Fremont, Ohio) and retailed by dental supply houses. After one min wipe gently with a cloth. By this treatment the constantan is stained red and the iron black. There is a sharp line of demarcation at the junction of the two metals. For copper and constantan, the original colors of the two metals suffice to distinguish them.

2. Lay the junction on a steel scale and with a sharp knife cut off to the specified length.

3. Square the end with a fine file and make the second weld in the same manner as the first.

The distance between the two junctions can be measured on the junction gage (see Fig. 5·8) in the following manner.

1. Clean the wire by rubbing with fine emery paper.
2. Stain with copper cement, as described above.
3. Lay on top of the junction gage approximately normal to the inclined straightedges. Have good illumination from above. Slide back and forth until a position is found where each junction is aligned with one of the straightedges.

4. Read the distance s (see Fig. 5·9) between junctions from the scale on the junction gage.

For 0.010-in. diameter wires the distance s can be produced to specifications to within ±0.005 in. and can be measured after completion to within ±0.001 in.

Butted junctions can be made in 0.0025 in. and smaller sizes of wire by electroplating. The location of the junction along the length of wire is not, however, definite by this method.[36]

Junctions can be made by soldering or brazing the twisted ends of the two thermocouple wires directly to the surface of the mass of metal whose temperature is to be measured. The methods described above can be applied by using a flux suited to the particular metal.[31] If desirable, the wires can be soldered or brazed individually, or together, in holes drilled in the metal (see Sec. 8·6).

After beading, or as individuals, the thermocouple wires can be butt-welded by the *electric-resistance-pulsation method* directly to the surface of the parent metal. Complete package equipment for this operation is available commercially.[31] This method cannot be satisfactorily utilized to weld wires to the bottoms of deep holes. Although strong welds can be made to steel and to most other metals, welds made to aluminum and aluminum alloys are of unsatisfactory strength (see Sec. 8·4).[37]

By the *condenser-discharge method*, beaded junctions or individual wires can be welded to the surface or to the bottom of a hole in the parent metal. Such welds are usually brittle, weak, and unreliable. They should not be attempted on aluminum or aluminum alloys. If it is necessary to perform such a weld, a beaded junction should be made first by one of the methods described above. Then, having connected the metallic mass to the positive end of the charged-condenser bank, and having gripped the thermocouple wires close to the bead with pliers, press the bead quickly against the metal block. The energy of the discharge at the instant of contact melts and partly vaporizes portions of both block and bead, and a weld is formed. In welding No. 30 B & S gage (0.010 in.) iron and constantan wire the best result is obtained by charging 1000 μf to 250 v d-c (see Fig. 5·7). Complete package equipment is available commercially.[31, 37]

Gas- or arc-welding of thermocouple wires to the parent metal can be done by machining on the latter a teat, flange, or other projection of approximately the same cross-sectional area as that of the wire. Welding to this projection is then performed as when joining two wires in the manner described above.

Splices in lead wires can be made by essentially the same procedure as with beaded junctions. Here, however, as indicated in Fig. 5·10,

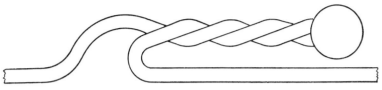

Fig. 5·10. Fused splice.

it is not necessary that all the twisted portion be melted into the bead. The two ends can be pulled out straight as shown. Insulating tape can be wound around the joint.

5·9 INSULATION

At elevated temperatures, where there is less danger of voltaic effects, the requirements for electrical insulation are less severe. Circumstances often make it difficult to attain fully effective insulation in the immediate vicinity of the "hot" junction. In designing an installation it is

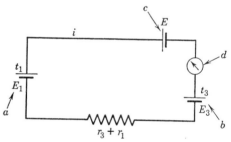

Fig. 5·11. Thermocouple circuit without insulation leakage. a, "hot" junction; b, "cold" junction; c, balancing emf; d, current indicator.

therefore desirable to know approximately what quality of insulation must be provided in order to attain a given degree of precision in temperature measurement.

Figure 5·11 shows the diagram for a simple thermocouple circuit. The "hot" and "cold" junctions at temperatures t_1 and t_3 are indicated as emfs, E_1 and E_3, respectively. i is the electric current, made as nearly as feasible equal to zero by the balancing emf, E, applied at the potentiometer; and r_3 and r_1 are electrical resistances in the circuit.

Figures 5·12, 13, 14 show this same circuit with *leakage* caused by

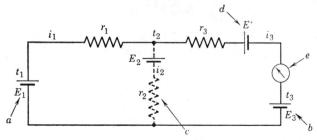

Fig. 5·12. Circuit of Fig. 5·11, with insulation leakage between leads. *a*, "hot" junction; *b*, "cold" junction; *c*, assumed insulation leakage; *d*, balancing emf; *e*, current indicator.

$$\Delta t_x/\Delta t_y = (t_1 - t_2)/(t_1 - t_3)$$

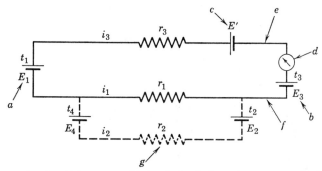

Fig. 5·13. Circuit of Fig. 5·11, with insulation leakage between one lead and the parent metal at two points. *a*, "hot" junction; *b*, "cold" junction; *c*, balancing emf; *d*, current indicator; *e*, metal A; *f*, metal B; *g*, assumed insulation leakage through metal C, i.e., parent metal.

$$\Delta t_x/\Delta t_y = (t_4 - t_2)/(t_1 - t_3)$$

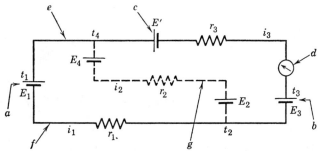

Fig. 5·14. Circuit of Fig. 5·11, with insulation leakage between both leads and the parent metal. *a*, "hot" junction; *b*, "cold" junction; *c*, balancing emf; *d*, current indicator; *e*, metal A; *f*, metal B; *g*, assumed insulation leakage through metal C, i.e., parent metal.

$$\Delta t_x/\Delta t_y = (t_1 - t_2)/(t_1 - t_3)$$

Sec. 5·9 INSULATION

an *insulation failure* of electrical resistance r_2. As a consequence of this partial *short* and the parasitic thermoelectric emfs introduced by thus including external metal in the circuit, the emf indicated on the potentiometer is now E' instead of E for the same temperatures, t_1 and t_3, of the junctions. The ratio, $(E - E')/E \times 100$, gives the resulting

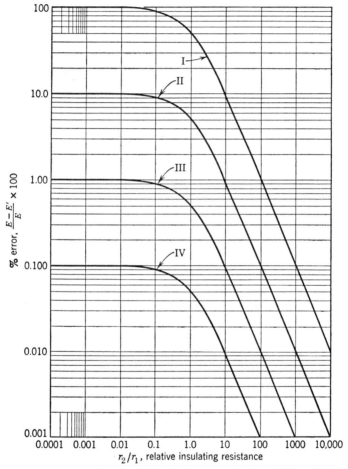

Fig. 5·15. Error curves for insulation leakages in circuits of Figs. 5·11, 12, 13, and 14.

Curve	$\Delta t_x / \Delta t_y$
I	1.000
II	0.100
III	0.010
IV	0.001

per cent of error in determination of temperature. The values of this error for various insulation resistances at the partial failure are plotted in Fig. 5·15.

5·10 CIRCUITS

Figure 5·16 shows a common circuit for reading a number of thermocouples with one *emf-indicating instrument* and one *ice point*. In this

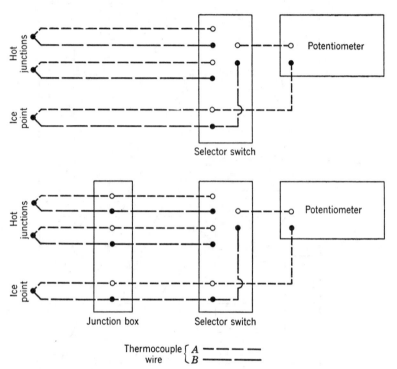

Fig. 5·16. Circuit for reading a number of thermocouples with one potentiometer and one ice point.

circuit, the thermocouple wires lead to the terminals at the *junction box*, the *selector switch*, and the *potentiometer*. It is therefore important that uniform temperatures be maintained throughout these instruments respectively. The selector switches and *potentiometer indicators* are usually built with good thermal communication between the various terminals. The junction box should also be so made. An added necessary precaution, except in low precision work, is to insulate these three instruments thermally. In precision work, the junctions with the copper leads are made individually at the ice point for each thermocouple.[1, 2, 10, 30, 38]

Figure 5·17 shows the circuit called a *thermopile*. The result of this arrangement is to multiply effectively the thermoelectric power by the

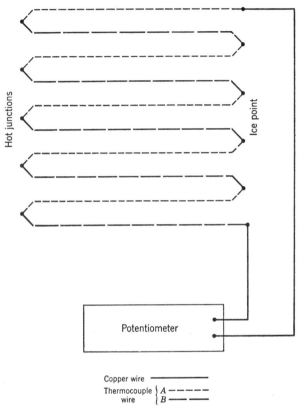

Fig. 5·17. Thermopile circuit.

number of couples thus placed electrically in series. Greater sensitivity can thus be attained. It is also the most reliable thermoelectric method for measuring the average, i.e., arithmetic mean, of the temperatures at a number of points. Connecting thermocouples in parallel for this purpose dispenses with the need for insulating the hot junctions from one another; however, the accuracy of the average thus determined is dubious.[33, 34, 39]

Figure 5·18 shows a well-designed ice point. An ordinary commercial Dewar flask can be used. Except in very high precision work, i.e., to ±0.0005°F, commercial ice cracked to "pea" size is satisfactory. Water should be just sufficient to fill the interstices between the pieces of ice. It is best to have the bottom of the tube, containing the junctions, 2 or 3 in. above the bottom of the ice space. Acid-free and moisture-free kerosene should be used. Moisture-proof insulation should extend to beneath the surface of the kerosene. It is necessary that a sufficient length of bare or lacquered wire be immersed in the kerosene to insure that the junction will be at the kerosene temperature. The length required depends on the diameter and thermal conductivity of wire used. The copper strand, being of high-thermal-conductivity material, usually causes the most trouble and should be No. 30 B & S gage (0.010 in.) or smaller. Larger-diameter copper wire can be joined to this a few feet from the ice point if so desired. The length of immersion should always be at least 6 in. Similarly, the glass tube containing the junctions should be immersed at least 6 in. in the ice bath.

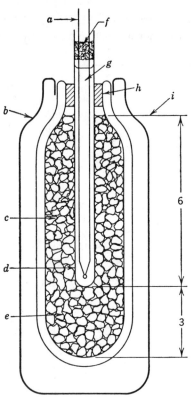

Fig. 5·18. Ice point. a, No. 30 B & S gage (0.010 in.) wire with moisture-resistant insulating coating extending below the surface of kerosene; b, Dewar flask; c, clean, pea-size ice extending to the bottom; d, ¼-in. diameter glass tube; e, water to fill interstices between ice granules; f, cotton plug; g, acid-free and moisture-free kerosene; h, rubber stopper; i, water level.

It is possible to put more than one tube in a Dewar flask, and, if lacquering provides adequate electrical insulation, more than one junction can be placed in a tube. However, an excessive number should be avoided, since they will tend to prevent one another from attaining the ice-point temperature.

Thermocouples can be read by either of two alternative methods—i.e., by the *direct method* or by the *null method*. In the direct method it is the electric current flowing in the thermocouple circuit which is indicated by the instruments. The thermoelectric emf causing this current is inferred through Ohm's law, which states that the current is equal to the ratio of the emf to the electrical resistance in the circuit. In the null method, the thermoelectric emf is indicated directly. The direct method involves less instrumentation and is therefore cheaper; it also reduces the number of points at which parasitic thermoelectric and voltaic-effect errors can be introduced. The thermoelectric emf which corresponds to a given current depends on the electrical resistance of the circuit which, in itself, varies with temperatures. Where it is desired to measure only such small temperature differences as permit using the full sensitivity of a sensitive *current indicator*, the direct method can, in general, be made the more precise, since the sensitivity is limited by that of the current indicator alone and not also by that of other instruments. For this reason the direct method is much used in laboratories on precision scientific work. It is rarely used in industrial test work. Thermoelectric emfs determined by the null method are independent of the circuit resistance and can be used directly with the standard tables for the kind of thermocouple wires used.[9, 40]

REFERENCES

1. W. F. Roeser, "Thermoelectric Thermometry," American Institute of Physics, *Temperature*, pp. 180–205, Reinhold Publishing Corp., New York, 1941.
2. P. Davidson, "Selection of Lead Wires for Base-Metal Thermocouples," *Combustion*, 14, no. 10, pp. 42–43 (April, 1943).
3. W. F. Roeser and A. I. Dahl, "Reference Tables for Iron-Constantan, and Copper-Constantan Thermocouples," RP 1080, *J. Research the Natl. Bur. Standards*, 20, no. 3, pp. 337–355 (March, 1938).
4. H. Shenker, J. I. Lauritzen, Jr., and R. J. Corruccini, "Reference Tables for Thermocouples," *Natl. Bur. Standards Circ.* 508, 71 pp., Government Printing Office, Washington (1951).
5. D. K. Warner, "The First Symposium on Standardization of Thermocouple Wire Specifications," *Instruments*, 24, no. 8, pp. 877, 957–958, 960 (August, 1951).
6. American Society of Mechanical Engineers, *A.S.M.E. Mechanical Catalog and Directory*, 42, p. 632, New York (1953).
7. A. Schulze, "Metallic Materials for Thermocouples," *J. Inst. Fuel*, 12, no. 64, pp. S41–S48 (March, 1939).
8. A. I. Dahl, "The Stability of Base-metal Thermocouples in Air from 800 to 2200°F.," American Institute of Physics, *Temperature*, pp. 1238–1266, Reinhold Publishing Corp., New York, 1941.
9. J. G. Aston, "The Use of Copper-Constantan Thermocouples for Measurement of Low Temperatures Particularly in Calorimetry," American Institute of Physics, *Temperature*, pp. 219–227, Reinhold Publishing Corp., New York, 1941.

10. D. D. Wile, "Measurement of Temperature in the Laboratory by Means of Thermocouples," *Refrig. Eng.*, Journal of the American Society of Refrigerating Engineers, **49**, no. 1, pp. 35–39 (January, 1945), and **49**, no. 2, pp. 118–120 (February, 1945).
11. W. F. Roeser, A. I. Dahl, and G. J. Gowens, "Standard Tables for Chromel-Alumel Thermocouples," RP 767, *J. Research Natl. Bur. Standards*, **14**, no. 3, pp. 242–243 (March, 1935).
12. W. I. Pumphrey, "The Embrittlement of Chromel and Alumel Thermocouple Wires," *J. Iron Steel Inst.*, **157**, Part 4, pp. 513–514 (December, 1947).
13. C. F. Homewood, "Factors Affecting the Life of Platinum Thermocouples," American Institute of Physics, *Temperature*, pp. 1272–1280, Reinhold Publishing Corp., New York, 1941.
14. B. Brenner, "Changes in Platinum Thermocouples Due to Oxidation," American Institute of Physics, *Temperature*, pp. 1281–1283, Reinhold Publishing Corp., New York, 1941.
15. The Liquid Steel Temperature Subcommittee, "A Symposium on the Contamination of Platinum Thermocouples," *J. Iron Steel Inst.*, **155**, Part 2, pp. 213–234 (February, 1947).
16. A. B. Kaufman, "Bismuth Thermocouples," *Instruments*, **25**, no. 6, pp. 762, 802–803 (June, 1952).
17. A. Schulze, *Metallische Werkstoffe für Thermoelemente* (Metallic Materials for Thermocouples), pp. 72–80, J. W. Edwards, Ann Arbor, 1946.
18. W. C. Troy and G. Steven, "High Temperature Thermocouple—New Metal Combination Measures up to 2200°C," *The Frontier*, **12**, no. 4, pp. 6–8, 22–24 (December, 1949).
19. B. Kinkul'kin, "New Thermoelectric Pyrometers" (in Russian), *Stal*, no. 6, pp. 15–19 (1938).
20. Engineering Societies Library, "Pyrometry and Thermoelectricity; with particular Reference to Tantalum, Tungsten, and Molybdenum, 1920–1945," *Search* 5166, 46 pp. (Typewritten), Engineering Societies Library, New York, 1945.
21. F. H. Morgan and W. E. Danforth, "Thermocouples of the Refractory Metals," *J. Appl. Phys.*, **21**, no. 2, pp. 112–113 (February, 1950).
22. G. R. Fitterer, "Thermoelectric Apparatus," *U. S. Patent Office* 2094102, 3 pp., Government Printing Office, Washington (1937).
23. R. R. Ridgway, "Thermocouple," *U. S. Patent Office* 2152153, 3 pp., Government Printing Office, Washington (1935).
24. W. F. Roeser and H. T. Wensel, "Methods of Testing Thermocouple Materials," American Institute of Physics, *Temperature*, pp. 284–314, Reinhold Publishing Corp., New York, 1941.
25. C. T. Weller, "Characteristics of Thermocouples," *Gen. Elec. Rev.*, **49**, no. 11, pp. 51–53 (November, 1946).
26. "Testing by the National Bureau of Standards," *Natl. Bur. Standards Circ.* 483, pp. 52–53, Government Printing Office, Washington (1949).
27. H. C. Quigley, "Résumé of Thermocouple Checking Procedures," *Instruments*, **25**, no. 5, p. 617 (May, 1952).
28. "Fused Stabilized Zirconia," *Mech. Eng.*, **73**, no. 6, pp. 507–508 (June, 1951).
29. American Society of Mechanical Engineers, pp. 384, 477, 481, 484, 527, 576, and 632 (see Ref. 6).

REFERENCES

30. Y. V. Baimakoff, "Resistance of Contact Between Metals, and Between Metals and Carbon Materials," *The Engineers' Digest*, 4, no. 4, p. 164 (April, 1947).
31. American Society of Mechanical Engineers, pp. 461, 477, 604, 660–662 (see Ref. 6).
32. R. W. Mebs and W. F. Roeser, "Solders and Soldering," *Natl. Bur. Standards Circ.* 492, pp. 1–12, Government Printing Office, Washington (1950).
33. N. P. Bailey, "The Response of Thermocouples," *Mech. Eng.*, 53, no. 11, pp. 797–804 (November, 1931).
34. H. Emmons, "The Theory and Application of Extended Surface Thermocouples," *J. Franklin Inst.*, 229, no. 1, pp. 29–52 (January, 1940).
35. E. F. Hammel, *Method for the Preparation of Butt-Welded Thermocouples Using 3-Mil Diameter Wire*, MDDC 776, LADC 393, pp. 1–6, U. S. Atomic Energy Commission, Oak Ridge, 1947.
36. H. J. Carter, "A Method of Constructing Cu-Constantan Thermocouples," *Rev. Sci. Instr.*, 19, no. 12, pp. 917–918 (December, 1948).
37. Trott, W. J., "Welder for Attaching Fine Wires to Massive Metal Bodies," *Rev. Sci. Instr.*, 20, no. 8, pp. 624–625 (August, 1949).
38. J. T. Cataldo, "Apparatus and Procedure for Testing Pyrometer Switches," *Instruments*, 21, no. 11, pp. 1014–1015 (November, 1948).
39. A. I. Dahl and E. F. Fiock, "The Use of Parallel Thermocouples in Turbojet Engines," *NBS Report* 1099, *USAF Technical Report* 6546, pp. 1–11, U. S. Air Force, Wright Air Development Center, Dayton (1951).
40. P. D. Foote, C. O. Fairchild, and T. R. Harrison, "Pyrometric Practice," *Technological Papers of the Bureau of Standards* 170, pp. 20–94, 189–209, 226–257, Government Printing Office, Washington (1921).

6

INDICATING INSTRUMENTS

6·1 SOURCES OF INSTRUMENTS

Progress in the field of *indicating instruments* is rapid at the present time. New designs are described in each monthly issue of each of the several magazines devoted to this subject. A large number of companies are engaged in the manufacture of instruments, which are variously protected by their patents. Lists of such companies are available in the indexes and buying guides and in the announcements and advertisements in the instrument journals.[1-7] Up-to-date and complete information respecting the products of any one manufacturer is only to be obtained through consulting his latest bulletins, supplemented by correspondence and, preferably, direct contact.

6·2 THEORY

The class of equipment categorized as indicating instruments provides not merely for indication of thermocouples and other temperature-sensitive elements, but also for elements sensitive to pressure, flow, torque, strain, and other phenomena. If the sensitive element is such as to yield the same type of *signal*, the same instrument can be used for indication, regardless of whether it is temperature, pressure, flow, or what not, that is being measured. Different meanings, merely, are associated with the dial readings. Thus, indication is a field in itself and not a subdivision of the field of temperature measurement. Furthermore, indication alone is but a special case. Recording, telemetering, and control are other divisions of the broader field. The general body of theory underlying this class of instrumentation technique is correspondingly extensive and treated in its special aspects by various excellent works.[2, 7-12]

Whereas the broader phases of the subject are thus beyond the scope of the present work, presentation of certain elements may effect a degree of orientation.

6·3 THE SIGNAL

Thermoelectric elements provide an electromotive force or emf which tends to drive an electric current in the thermocouple circuit. This emf, or signal, is small, ranging from 5 to 35 μv (microvolts) per °F for the usual materials. Thus, measurement of temperature to within 1°F requires measuring emf accurately to within 0.5 to 3.5 μv (see Eq. 3·4), whereas measurement of temperature to within 0.0001°F requires measurement of emf to within 0.00005 to 0.00035 μv. These emfs are of the order of magnitude of those of the Johnson noise. Use of multiple junctions increases the magnitude of emf to be measured by a factor equal to the number of junctions in series.

6·4 PARASITIC EMFS

Not only do these small magnitudes of emf demand sensitive types of indicating instruments, but they render it imperative that parasitic or spurious sources of emf be reduced to a minimum. Thus, it is essential that all auxiliary instruments be constructed so that they will not, in themselves, introduce parasitic emfs and, preferably, will be so adapted as to minimize the effects of those unavoidably introduced elsewhere in the circuit.[8]

6·5 EMF VS. CURRENT

The signal is thus initially in the form of a small emf. However, by Ohm's law

$$i = E/r \qquad (6·1)$$

where E is the thermoelectric emf in volts, r is the circuit resistance in ohms, and i is the electric current intensity in amperes. This signal may thereby also be regarded as a small electric current.

6·6 TYPES OF INDICATORS

Indicating instruments for use with thermocouples fall into two classes: emf indicators and current indicators. Emf indication possesses the advantage of eliminating the variable r, which is usually a function of the ambient temperature. Current indication is, on the other hand, disposed to be somewhat simpler, with the effects of greater ruggedness and convenience in reading or sensitivity for a given expenditure in instrumentation. For example, current indication can be effected by including a galvanometer element in the circuit. This galvanometer then functions essentially as a microammeter or millimicroammeter. Emf indication, on the other hand, is usually performed by comparing the thermoelectric emf with a reference emf, such as that of a standard-type voltaic cell. Such comparison requires a balancing operation, similar to that of weighing on a scale, using a current indicator to reveal the residual degree of unbalance in the course of successive adjustments. The balancing circuit, analogous to the weighing scale, is called a *potentiometer*.[8] The introduction of a potentiometer into the circuit usually lowers the sensitivity of a given current indicator with respect to the objective of measuring thermoelectric emf. A standard cell, properly handled, is highly stable, whereas most current indicators tend to drift, necessitating frequent recalibration. Introduction of additional elements, such as the potentiometer, standard cell, and switching devices, tends to introduce additional parasitic emfs, necessitating additional precautions. The balancing operation tends to be time-consuming and a source of personal error. However, apparatus is available for performing this operation automatically, effecting direct-reading indication, recording, telemetering, or control actuation. Introduction of these automatic arrangements usually results in lowered precision.

6·7 ELECTRONIC AMPLIFIERS VS. SUSPENSION GALVANOMETERS

The thermoelectric emf or current is, in unmodified form, nonalternating, i.e., direct current, and is, as such, inconvenient to amplify by the more usual electronic techniques. Thus moving-coil and moving-magnet-type suspension galvanometers were long the principal means of indication. However, means of modulation have been developed to convert this d-c to an a-c signal, whereby indicators are now available devoid of delicate mechanical elements.[13] In their most sensitive forms suspension galvanometers require the highest degree of isolation from vibration and other mechanical disturbances. Thus, the use of some form of Julius suspension[14] is necessary where stable conditions do not prevail. Even the nonprecision types require some protection. The completely electrical, i.e., *electronic*, indicators, in form developing sensitivities comparable to those of the most sensitive suspension galvanometers, are comparatively immune to the effects of mechanical disturbance. They are furnished in compact, self-contained packages, as are also nonprecision and semiprecision moving-coil systems. However, the most sensitive, precision, suspension galvanometers require cumbersome, auxiliary optical systems. Drift and servicing problems are more acute and prevalent with the electronic units. The ultimate sensitivity in both the electronic and suspension galvanometer systems is the same, both being limited by the Johnson noise.

6·8 PRECISION REQUIREMENTS

It is indispensable that instrumentation selected for the execution of a given measurement be adequate to satisfy the precision requirements for that job (see Secs. 3·1, 7). From among alternative arrangements, each satisfying this fundamental condition, consideration should be given to matters of expense, ruggedness, convenience in reading, adaptability to semiskilled labor, etc., in the order in which these factors attain importance in the particular situation.[15]

REFERENCES

1. "Apparatus Directory," *J. Appl. Phys.*, **11**, no. 6, pp. iv–xx (June, 1940).
2. American Society of Mechanical Engineers, *Sources of Information on Instruments*, pp. 3–16, New York, 1945.
3. "The 1950 Instruments Index," *Instruments*, Part 2, pp. 20–136 (July, 1949).
4. *Radio's Master*, 16th edition, pp. F 1–6, 11, 15, 18–20, 23–25, 33–38, 48, 51–60, 64–67, 78, 95, 98, 99, 102, 103, 105, 108, L 1–4, 18–30, M 1–14, United Catalog Publishers, New York (1951).
5. American Society of Mechanical Engineers, *A.S.M.E. Mechanical Catalog and Directory*, **42**, pp. 378, 433, 435, 484–487, 515–517, 532, 534, 559, 573, 576–580, and 632, New York (1953).
6. *Industrial Laboratories*, Industrial Laboratories Publishing Co., Chicago.
7. *Rev. Sci. Instr.*, American Institute of Physics, New York.
8. W. P. White, "Potentiometers for Thermoelectric Measurements," American Institute of Physics, *Temperature*, pp. 265–278, Reinhold Publishing Corp., New York (1941).
9. G. S. Brown and D. P. Campbell, *Principles of Servomechanisms*, 400 pp., John Wiley & Sons, New York, 1948.
10. D. P. Eckman, *Industrial Instrumentation*, 396 pp., John Wiley & Sons, New York, 1950.
11. H. Chestnut and R. W. Mayer, *Servomechanisms and Regulating System Design*, **I**, 505 pp., John Wiley & Sons, New York, 1951.
12. G. H. Farrington, *Fundamentals of Automatic Control*, 285 pp., John Wiley & Sons, New York, 1951.
13. M. D. Liston, C. E. Quinn, W. E. Sargeant, and G. G. Scott, "A Contact Modulated Amplifier to Replace Sensitive Suspension Galvanometers," *Rev. Sci. Instr.*, **17**, no. 5, pp. 194–198 (May 1, 1946).
14. H. D. Baker, W. Claypoole, and D. D. Fuller, "Further Developments in the Measurement of the Coefficient of Static Friction," *Proceedings of First U. S. National Congress of Applied Mechanics*, American Society of Mechanical Engineers, New York (1951).
15. British Standards Institution, "Temperature Measurement," *British Standard Code*, B.S. 1041: 1943, pp. 41–48, London (1943).

7

DESIGN CALCULATION TECHNIQUES

7·1 TEMPERATURE AT A POINT

The temperature of a solid body may vary from point to point within the extent of the body and at any one point may vary with time. Accuracy in measurement requires accuracy not only in temperature measurement itself but also in the measurement of the time and location at which the temperature occurs.

7·2 THERMAL CONTACT

To insure that the temperature measured corresponds to the actual temperature of the body, thermal *contact* must be established between the sensitive element and the body. The thermal resistance R_1 at this contact must be sufficiently small as compared to the thermal resistance R_2 between the sensitive element and the external surroundings or *ambient*. If this condition is met, the temperature assumed by the sensitive element will correspond to that of the body to within the required degree of precision. Thus, the sensitive element (see Fig. 7·1)

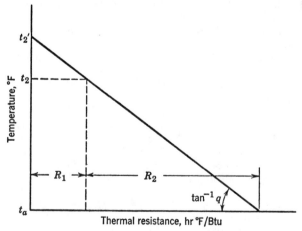

Fig. 7·1. Temperature vs. thermal-resistance diagram.

assumes a temperature t_2 intermediate between the temperature of the body at the specified point of measurement t_2' and the temperature of the surroundings t_a. This temperature assumed by the sensitive element varies in accord with the ratio of the above two resistances R_1 and R_2.

7·3 NONUNIFORM TEMPERATURE

If the temperature of the body is nonuniform—in particular, if substantial variations in temperature occur in very short intervals of distance within the body—a suitable installation design must be employed to assure the measurement of local temperatures, with an adequate degree of precision. Consideration must be given to possible alteration of the temperature distribution in the body and to corresponding errors in the temperature, resulting from removal of material to provide space for insertion of the sensitive element. Installation of the sensitive element can be utilized partially to restore the heat-flow paths cut by the removal of materials, with a corresponding decrease in error. However, the smaller the amount of material removed (especially in the vicinity of the point at which it is desired to measure temperature), the less will be the resultant disturbance in temperature distribution, with a correspondingly smaller error in the temperature measured.

7·4 INCONSTANT TEMPERATURE

If the temperature of the body varies with time at the point at which temperature is to be measured, it is essential that the temperature of the sensitive element reflect these temperature changes of the body. The external instrumentation must then also respond with sufficient rapidity to follow the temperature changes of the sensitive element. The time interval by which the temperature of the sensitive element lags behind that of the adjacent portion of the body will be in proportion to the product of the thermal capacity of the element and the thermal resistance, R_1, of the heat-flow path from the adjacent parent material to the vital zone within the sensitive element. This product will be less for small, sensitive elements made of material of high thermal conductivity and for installations providing small thermal resistance, R_1, between the sensitive element and the adjacent material.

7·5 HEAT TRANSFER

Referring to Fig. 7·1, if R_2 is small, q may be substantial. Then the rate of heat flow q to or from the point in the body at which it is desired to measure temperature may be such as to result in appreciable local disturbance in the temperature of the body. R_1 must then be regarded as the thermal resistance between the sensitive element and points of the body outside this region of local disturbance. R_1 will depend on the thermal conductivity of the material of which the body is composed, the size of the excavation, and the degree of thermal contact between the sensitive element and the adjacent points in the parent body. Then, for a given value of R_1, the value of R_2 must be increased to such a magnitude as is necessary to bring the error in temperature measurement, $t_2' - t_2$, to within acceptable limits. Thus, an increase in R_2 effects a decrease in q, which, in turn, results in a decrease in $t_2' - t_2$.

7·6 INSTALLATION DESIGN CONSIDERATIONS

R_2 can be increased (1), by making the leads of minimum cross-sectional area; (2), by making the leads of material of minimum thermal conductivity; (3), by thermally insulating the external leads and, (4), by providing a sufficient length in the leads to the first point at which good thermal contact occurs with the ambient at temperature t_a. Such a point may be at a binding post on an auxiliary instrument or at a cold junction.

Assuming the external leads are insulated and of adequate length, the conditions that make for the highest degree of precision in an installation are (1), maximum thermal contact between the sensitive element and the adjacent portions of the body; (2), minimum dimensions in the sensitive element; (3), minimum metallic cross-sectional area in the leads; and, (4), minimum thermal conductivity in the materials of which the leads are composed.

Achievement of conditions (1) and (4) above is desirable in all installations but is limited by the possibilities of actual methods and materials. Achievement of conditions (2) and (3) can be attained only at the expense of ruggedness. Thus, smaller sizes of wire result in greater precision but are more fragile. Difficulties in drilling the holes and in handling the parts during preparation and installation increase as the size decreases. Hence, in the interests of minimum cost and maximum ruggedness, the largest size of wire is used which will permit satisfaction of the precision requirements. Therefore, installation designs aimed at the achievement of conditions (1) and (4) should be properly considered as occurring in families in which the individual members differ essentially in wire size, i.e., in conditions (2) and (3) above.

7·7 ENGINEERING PRACTICE IN DESIGN

In engineering practice, it is customary to base design on quantitative calculations from established facts, as supplemented by judgment. The effects of errors: (1), in values assumed for properties of materials; (2), in calculation techniques employed; (3), in estimations by judgment; and, (4), in expected operating conditions, are intended to be provided for by a *factor of safety*. The factor of safety is to be so chosen as to provide a margin, such that, allowing for all possible discrepancies, it is still assured that the performance of the device will conform to requirements.

7·8 DESIGN FOR TEMPERATURE MEASUREMENT

In the field of temperature measurement, the procedure of Sec. 7·7 has all too frequently been curtailed to the exercise of judgment as tempered by optimism. Both the design calculations and the factor of safety have been omitted. This is often extremely unfortunate; as, except by means of an extended test program (rarely executed), it is usually difficult or impossible to determine the actual errors to which installations are subject. The resultant effect has often been for temperature measurement data to become a source of misinformation. It is the purpose of the following sections to present calculation techniques that are simple enough for convenient application yet sufficiently accurate if used with the usual engineering factor of safety in design. Further elaborations in calculation technique would normally add more to the expense of the design work than is justified.

7·9 HEAT TRANSFER TO THE AMBIENT VIA THE LEADS

Usually, the leads assume a temperature, roughly approximating that of the body in which the installation is made, at their point of entry into this body. Likewise, in most actual arrangements of apparatus, sufficient length of leads exists for attainment of an approximation to the ambient temperature, before reaching the point of contact with the external instrumentation. Under these idealized circumstances, the hourly heat, q (see Fig. 7·2), transferred from the body to the ambient via the installation, is given by

$$q = (UCkA)^{1/2}(t_1' - t_a) \qquad (7 \cdot 1)$$

where q is the heat transferred, Btu/hr; C is the circumference of the leads, ft; U is the effective value of the surface boundary conductance as referred to the circumference C of the leads, Btu/hr ft² °F; k is the mean value of the thermal conductivity of the lead materials, Btu/hr ft °F; A is the cross-sectional area of the leads, ft²; t_1' is the surface temperature of the body, °F; and t_a is the ambient temperature, °F.

Where the leads are metal thermocouple wires covered with electrical insulation, it is usually of sufficient accuracy to take A as the sum of the two wire cross sections and k as the average of the thermal conductivities of the two metals.

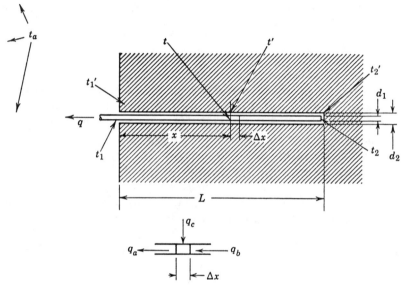

Fig. 7·2. Installation, temperature-pattern, and heat-flow diagram.

U, as referred to the circumference C of the electrical insulation, is given by

$$U = \frac{1}{1/(U_f^o + U_r^o) + N_1/U_w + N_2/(U_f^i/2 + U_r^i)} \quad (7.2)$$

where U_f^o is the fluid boundary conductance on the outside surface C, Btu/hr ft² °F; U_r^o is the radiation boundary conductance at the outside surface C, Btu/hr ft² °F; U_w is the average thermal conductance for one layer of electrical insulating material referred to circumference C, Btu/hr ft² °F; U_f^i is the fluid boundary conductance on the internal surfaces as referred to circumference C, Btu/hr ft² °F; U_r^i is the radiation boundary conductance for exchange between the internal surfaces as referred to circumference C, Btu/hr ft² °F; N_1 is the number

Sec. 7·9 HEAT TRANSFER TO THE AMBIENT VIA THE LEADS

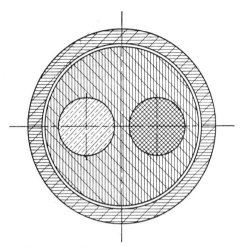

Fig. 7·3. Multiple electrical insulation.

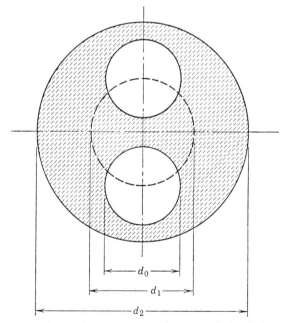

Fig. 7·4. Effective inner diameter of a two-hole insulating tube.

of air-separated layers of solid electrical insulating material; and N_2 is the number of pairs of opposing, internal, air-solid surfaces occurring in the wrapping.[1-3]

If the electrical insulation consists of one solid layer in intimate contact with the metal wires, $N_1 = 1$ and $N_2 = 0$. If a loose "spaghetti," protecting tube, or sheathing surrounds this insulated wire (see Figs. 7·3, 7·4), $N_1 = 2$ and $N_2 = 1$.

For any one solid insulating layer, U_w is given by

$$U_w^- = \frac{2\pi k}{C \log d_2/d_1} \qquad (7\cdot3)$$

where d_1 is the *effective* inside diameter of the layer, ft; d_2 is the external diameter of the layer, ft; and k is the mean thermal conductivity of the insulating material, Btu/hr ft °F.[1-3] For convenience in computations, this relationship is also plotted graphically in Fig. 7·5.

If the layer is a one-holed concentric tube, d_1 is simply its inside diameter. However, if the layer is a two-holed tube (see Fig. 7·5), d_1 is usually given with sufficient accuracy by

$$d_1 = \sqrt{2}\, d_0 \qquad (7\cdot4)$$

where d_0 is the diameter of the individual holes, ft. If the insulating layer is a cylindrical molding, surrounding and in contact with the pair of thermocouple wires, d_0 is the diameter of the individual wires, ft.

Values for U_f^o, U_r^o, U_f^i, U_r^i, and k can be obtained by referring to any standard work on heat transfer.[1-3]

7·10 SURFACE INSTALLATION

A thermocouple junction may make contact with the surface of a body at temperature t_1', transferring heat q via the leads to the ambient. This corresponds to the example of $L = 0$ in Fig. 7·2. Such a junction may be soldered, brazed, or welded to the surface (see Fig. 8·17), whereupon the thermal resistance at the contact is assumed to be zero. Again, it may be pressed against the bare surface or separated from it by a layer of electrical insulation or cement.

Then, the error in temperature measurement caused by the disturbance in local temperature by the heat efflux q is given by

$$t_1' - t_1 = q/\pi k D + q/A U_s \qquad (7\cdot5)$$

Sec. 7·10 SURFACE INSTALLATION 75

where q is given by Eq. 7·1, Btu/hr; k is the thermal conductivity of the parent-body material, Btu/hr ft °F; D is the diameter (D in Fig. 5·5) of the sensitive element, ft; U_s is the surface boundary conduc-

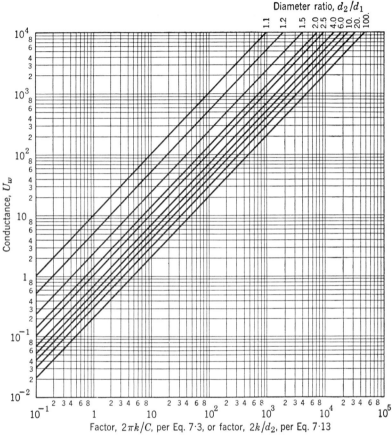

Fig. 7·5. Thermal conductance of electrical insulation per Eqs. 7·3, 13.

tance over the contact area, A, between the sensitive element and the parent-body material, Btu/hr ft² °F; and A is contact area, ft².

For soldered, brazed, or welded contact, the second term in Eq. 7·5, i.e., q/AU_s, may be neglected. For smooth, bare metal tightly pressed against smooth, bare metal, values for U_s can be estimated from experimental data. Depending on the actual materials, the contact pressure, their degree of smoothness, the temperature, and the lubricant (if any), values for U_s, ranging from 150 to 15,000 Btu/hr ft² °F, have been observed.[4,5]

If the contact is loose, with an intervening layer of insulating material,

$$U_s = \frac{1}{2/(U_f/2 + U_r) + 1/U_w} \qquad (7 \cdot 6)$$

where U_f is the fluid boundary conductance at the internal surfaces as referred to area A, Btu/hr ft² °F; U_r is the radiation boundary conductance at the internal surfaces as referred to area A, Btu/hr ft² °F; and U_w is the thermal conductance of the layer of insulating material as referred to area A, Btu/hr ft² °F.

$$U_w = k/h \qquad (7 \cdot 7)$$

where k is the thermal conductivity of the electrical insulating material, Btu/hr ft °F; and h is the thickness of the layer of insulating material, ft.

If no insulating layer occurs, the second term in the denominator of Eq. 7·6 vanishes and the first term becomes $1/(U_f/2 + U_r)$. If the insulating layer is in intimate or "wetted" contact with one of the surfaces, this first term in the denominator is $1/(U_f/2 + U_r)$. If the insulating layer is in "wetted" contact with both surfaces, the first term in the denominator of Eq. 7·6 vanishes.

7·11 CONTACT AT AN INTERNAL POINT

If contact of the sort discussed in Sec. 7·10 occurs at an internal point in the body, with the length L (see Fig. 7·2) of the leads loose or insulated in the hole (see Secs. 8·3 to 8·9), Eq. 7·5 becomes

$$t_2' - t_2 = q/2\pi kD + q/AU_s \qquad (7 \cdot 8)$$

7·12 CONTACT ALONG THE LENGTH OF IMMERSION

The installation may be of the *cemented* type described in Secs. 8·6, 11·2, 11·5 to 11·8, and 12·2 to 12·4. Or it may be some other arrangement (see Secs. 8·3, 6) where the heat efflux q to the ambient is transferred from the body to the leads along the length L of immersion of the leads in the parent-body material (see Fig. 7·2) instead of occurring only at the junction, as assumed in Sec. 7·11. The heat-flow pattern for this case is indicated in Fig. 7·2. The error in temperature measurement, due to discrepancy in junction temperature, is given by [6]

Sec. 7·12 CONTACT ALONG THE LENGTH OF IMMERSION

$$t_2' - t_2 = \frac{q}{kAm} \operatorname{csch} mL + \frac{g}{m} \coth mL \qquad (7\cdot9)$$

where
$$m^2 = UC/kA \qquad (7\cdot10)$$

and
$$C = \pi d_2 \qquad (7\cdot11)$$

and L is the length of immersion, ft; q is the heat flow in the leads at their point of exit from the body as given by Eq. 7·1, Btu/hr; k is the mean thermal conductivity of the immersed leads, Btu/hr ft °F; A is the cross-sectional area of the immersed leads, ft²; d_2 (see Fig. 7·2) is the diameter of the immersed leads or of the hole in which the immersion occurs, ft; g is the rate of increase of temperature in the direction of increasing x in Fig. 7·2, i.e., g is the component in the direction of the leads of the temperature gradient in the body at the location of the junction (see Sec. 12·1), °F/ft; t_2 is the temperature of the junction, °F; t_2' is the temperature of the undisturbed parent-body material at the location of the junction, °F; and U is the effective value of the surface conductivity as referred to diameter d_2, Btu/hr ft² °F, and is given by

$$U = \frac{1}{2/(U_f/2 + U_r) + 1/U_w + 1/(U_b + U_{rb})} \qquad (7\cdot12)$$

For convenience in computations the relationship (Eq. 7·9) is plotted graphically in Fig. 7·6.

The term $2/(U_f/2 + U_r)$ in the denominator of Eq. 7·12 corresponds to the thermal resistance at the surfaces of any loose insulating tube, where U_f is the fluid boundary conductance on the internal surfaces as referred to diameter d_2, Btu/hr ft² °F; and U_r is the radiation boundary conductance at the internal surfaces as referred to diameter d_2, Btu/hr ft² °F. For the cemented-type installation described in Secs. 11·2; 11·5 to 11·8; and 12·2 to 12·4, where the hole is filled solidly with cement in intimate contact with both the thermocouple wires and the wall of the hole, this term is neglected. U_w is given by

$$U_w = \frac{2k}{d_2 \log(d_2/d_1)} \qquad (7\cdot13)$$

where U_w is the conductance of the layer of cement or other material filling the annular space between the thermocouple wires and the inner wall of the hole as referred to diameter d_2, Btu/hr ft² °F; k is the thermal conductivity of the cement or other filling material, Btu/hr

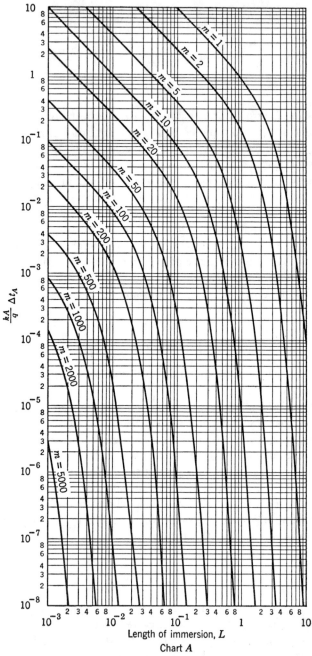

Fig. 7·6. Temperature error, Fig. 7·2, for immersed installation per Eq. 7·9.

$$t_2' - t_2 = \Delta t_A + \Delta t_B$$

Sec. 7·12 CONTACT ALONG THE LENGTH OF IMMERSION

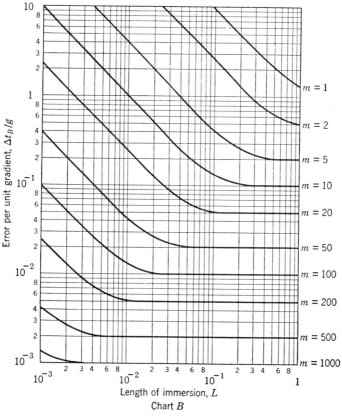

Chart B

Fig. 7·6. (*Continued*)

ft °F; d_2 is the diameter of the hole, ft; and d_1 (see Figs. 7·2, 4) is given by Eq. 7·4.[1-3] For convenience in computations, this relationship is also plotted graphically in Fig. 7·5.

The term $1/(U_b + U_{rb})$ in the denominator of Eq. 7·12 corresponds to the thermal resistance between unit area of the wall of the hole and portions of the body sufficiently distant not to be affected by the flow q of heat through the installation to the ambient. U_b can be computed with sufficient accuracy from the relation *

$$U_b = \frac{3k}{d_2 \sinh^{-1}(2L/d_2)} \tag{7·14}$$

where k is the thermal conductivity of the parent-body material, Btu/hr ft °F; d_2 is the diameter of the hole, ft; and L is the length of

* This is based on the temperature gradient laterally, from a conducting, prolate spheroid of semiaxes L and $d_2/2$, at a point $x = 3L/4$ (see Fig. 7·2).[7]

immersion, ft (see Fig. 7·2). For convenience in computations, this relationship is also plotted graphically in Fig. 7·7.

U_{rb} is the radiation boundary conductance as referred to diameter d_2, Btu/hr ft² °F. U_{rb} vanishes except with transparent bodies such

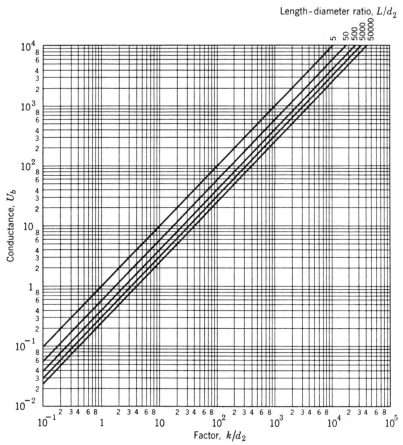

Fig. 7·7. Thermal boundary conductance of parent body per Eq. 7·14.

as glass. For such bodies it will be necessary to estimate values for U_{rb} from radiation-exchange-calculation technique, as described in detail in standard reference books on heat transfer.[1-3]

7·13 EXAMPLE

To illustrate the application of these computation techniques the following sample problem is worked out. It is supposed that a design

Sec. 7·13 EXAMPLE 81

as described in Sec. 11·2, Fig. 11·1, has been proposed for installation in a large steel machine part operating at a surface-temperature level estimated to be approximately 400°F. Operating conditions indicate the expectation of a thermal current flowing in a direction inclined at an angle of 45° to the axis of the proposed installation and of such magnitude as to effect a temperature gradient of approximately 175°F per in. in this direction. A study of the precision requirements in the project as a whole has indicated that the probable error P in this measurement must not exceed 5°F, i.e., allowable error $A = 3P = (3)(5) = 15°F$, see Eq. 3·1.

Thus it is required that the temperature error, $t_2' - t_2$, in Fig. 7·2, as evaluated for this case, be less than 5°F by a sufficient amount to provide for such errors as it may prove necessary or convenient to tolerate in the indicating instrumentation, and as due to deviations from the standard tables in the thermocouple wire and to parasitic thermoelectric and voltaic effects in the circuit.

The hourly heat, q, leaving the installation via the leads is first computed from Eq. 7·1. Thus

$$q = (UCkA)^{1/2}(t_1' - t_a) \tag{7·1}$$

where $t_1' = 400°F$, $t_a = 68°F$, $C = \pi(0.070)/12 = 0.0183$ ft (the outside diameter of the leads being taken as 0.070 in.), $k = 26$ Btu/hr ft °F (average thermal conductivity for iron and constantan),

$$A = \pi(0.010)^2/(2)(144) = 0.00000109 \text{ ft.}^2$$

U must be computed from Eq. 7·2. Thus

$$U = \frac{1}{1/(U_f^o + U_r^o) + N_1/U_w + N_2/(U_f^i/2 + U_r^i)} \tag{7·2}$$

where $N_1 = 2$, $N_2 = 1$, and where the surface conductances U_f^o, U_r^o, U_f^i, and U_r^i are estimated from data given in reference texts on heat transfer as $U_f^o + U_r^o = 5$ Btu/hr ft² °F and $U_f^i/2 + U_r^i = 4$ Btu/hr ft² °F.[1-3]

The thermal resistance of the electrical insulation on the leads must be determined from Eq. 7·3 or Fig. 7·5. Thus, $2\pi k/C = 2\pi(0.10)/(0.0183) = 34.3$ Btu/hr ft² °F (for lacquered, glass-fiber winding), $d_2/d_1 = 1.8$ (average ratio of outside to inside diameters for the inner windings and the external "spaghetti"), and U_w is read from the chart as 58 Btu/hr ft² °F.

Then, $U = 1/(\frac{1}{5} + \frac{2}{58} + \frac{1}{4}) = 2.06$ Btu/hr ft² °F, and $q = [(2.06)(0.0183)(26)(0.00000109)]^{\frac{1}{2}}(400 - 68) = 0.344$ Btu/hr.

In order to use Eq. 7·9 or the charts (Fig. 7·6), it is necessary first to compute m from Eq. 7·10. Thus

$$m^2 = UC/kA \qquad (7 \cdot 10)$$

where $C = \pi(0.046)/(12) = 0.012$ ft, $k = 26$ Btu/hr ft °F, $A = 0.00000109$ ft², and U must be computed from Eq. 7·12. Thus,

$$U = \frac{1}{2/(U_f/2 + U_r) + 1/U_w + 1/(U_b + U_{rb})} \qquad (7 \cdot 12)$$

where $2/(U_f/2 + U_r)$ and U_{rb} are both zero. U_w is determined from Eq. 7·13 or Fig. 7·5. Thus, $2k/d_2 = (2)(2)(12)/(0.046) = 1043$ Btu/hr ft² °F ($k = 2$ Btu/hr ft °F for the cement) and $d_2/d_1 = (0.046)/\sqrt{2}(0.010) = 3.25$. Then, U_w is read from the chart (Fig. 7·5), as 890 Btu/hr ft² °F. U_b is determined from Eq. 7·14 or Fig. 7·7. Thus, $k/d_2 = (26)(12)/(0.046) = 6780$ Btu/hr ft² °F (for steel), and $L/d_2 = (0.250)/(0.046) = 5.43$. Then, U_b is read from the chart as 6600 Btu/hr ft² °F, and $U = 1/(1/890 + 1/6600) = 785$ Btu/hr ft² °F.

Thus, $m = [(785)(0.012)/(26)(0.00000109)]^{\frac{1}{2}} = 577$ per ft.

From Chart A, Fig. 7·6, for $m = 577$ per ft and $L = 0.25/12 = 0.0208$ ft, $kA\Delta t_A/q$ is read as $(2.1)(10^{-8})$ ft, where $kA/q = (26)(0.00000109)/(0.344) = 0.0000824$ ft/°F. The result is, $\Delta t_A = (2.1)(10^{-8})/(0.0000824) = 0.00025$°F.

From Chart B, Fig. 7·6, $\Delta t_B/g$ is read as 0.00175 ft. The component of the temperature gradient in the direction of the axis of the installation is $(175)(12)/\sqrt{2} = 1485$ °F/ft. Δt_B is thus found to be $(1485)(0.00175) = 2.60$°F. The temperature-measurement error, $t_2' - t_2$ in Fig. 7·2, which it is desired to compute, is then $t_2' - t_2 = \Delta t_A + \Delta t_B = 0.00025 + 2.60 = 2.60025$°F.

Since the error, 2.6°F, is a systematic error, it cannot be reduced by repeated readings. It is also to be noted from Fig. 7·6 that this error will not be further decreased by increasing the length of immersion, L. It can, however, be decreased through reduction of m by using an installation such as shown in Figs. 11·7, 11·10, or 11·11.

Of the 5°F total permissible error, 2.4°F is thus available for other errors in the measurement problem.

If thermocouple wire conforming to the standard tables to within ¼ per cent is used, the error due to the wire will be approximately 1.0°F, leaving 1.4°F for all other errors. Indicating instrumentation will have to be selected in accord with this requirement and parasitic emfs reduced by exercising the recommended precautions.

REFERENCES

1. M. Jakob and G. A. Hawkins, *Elements of Heat Transfer and Insulation*, 169 pp., John Wiley & Sons, New York, 1942.
2. W. H. McAdams, *Heat Transmission*, 459 pp., McGraw-Hill Book Co., New York, 1942.
3. M. Jakob, *Heat Transfer*, I, 758 pp., John Wiley & Sons, New York, 1949.
4. N. D. Weills and E. A. Ryder, "Thermal Resistance Measurements of Joints Formed between Stationary Metal Surfaces," *Paper* 48-SA-43 *Trans. ASME*, 71, no. 3, p. 260 (April, 1949).
5. T. N. Centinkale and M. Fishenden, "Thermal Conductance of Metal Surfaces in Contact," *General Discussion on Heat Transfer*, London Conference, Sec. III, pp. 9–13, Institution of Mechanical Engineers, London (1951).
6. H. D. Baker and E. A. Ryder, "A Method of Measuring Local Internal Temperatures in Solids," *Paper* 50-A-101, pp. 1–2, American Society of Mechanical Engineers, New York (1950).
7. W. E. Byerly, *An Elementary Treatise on Fourier Series and Spherical, Cylindrical, and Ellipsoidal Harmonics*, p. 250, Ginn and Co., New York, 1893.

8

INSTALLATION DESIGN TYPES

8·1 SCOPE OF CHAPTER

Methods are given below for installing thermocouples in solids to measure interior temperature, including temperature at a point near the surface when approached internally, or even externally if the method is thermoelectric and similar to that required for internal approach. In temperature-measurement practice a number of different types of thermoelectric installation designs have been found useful under varying circumstances. This chapter presents one or more examples of each type.

8·2 PEENED JUNCTIONS

The method [1-6] of Fig. 8·1 [1,2] has been used in boiler tubes. Two holes are drilled near each other $\frac{1}{16}$ in. deep and 0.001 in. larger than

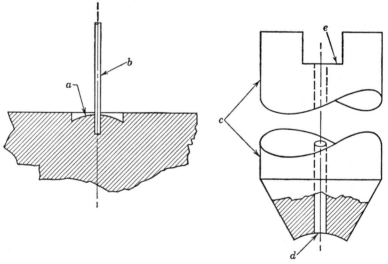

Fig. 8·1. Peened installation. *a*, indentation forcing metal around wire; *b*, thermocouple wire; *c*, peening tool; *d*, clearance hole for wire; *e*, groove for wire when hammering end of peening tool.

the wire diameter. The wires are separately inserted through the axial hole in the peening tool, which is then hammered lightly, the wires

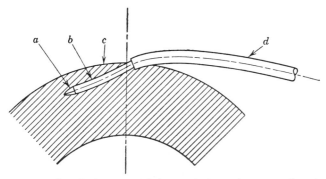

Fig. 8·2. Peened installation. *a*, chisel cut; *b*, bare thermocouple wire; *c*, lip forced back around wire; *d*, leads.

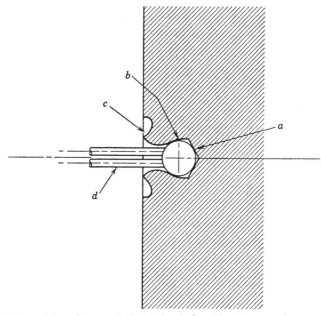

Fig. 8·3. Peened junction. *a*, drilled hole; *b*, fused bead; *c*, metal peened around bead; *d*, thermocouple wires.

being protected against injury by emergence in the crossgroove at the top. The junction is formed through the parent metal in which the wires are thus tightly embedded.

In Fig. 8·2,[7,8] also used for boiler tubes, a $\frac{1}{32}$-in. deep by $\frac{1}{8}$-in. wide lip is first gouged with a round-nosed chisel. This simplifies the drilling of two slanting holes $\frac{1}{10}$ in. deep and $\frac{1}{20}$ in. apart in which

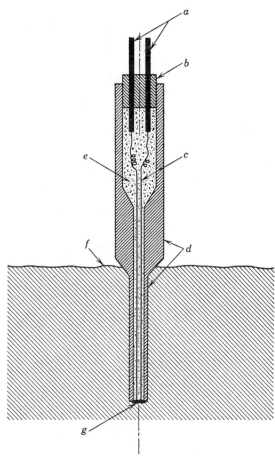

Fig. 8·4. Pierced-in junction; automobile-tire thermocouple. *a*, copper-constantan leads; *b*, maple plug; *c*, No. 36 B & S gage (0.005 in.) copper and constantan wire; *d*, maple tube; *e*, magnesium oxide powder; *f*, parent material (rubber tire); *g*, copper disk.

the separate wires are inserted. The lip is then peened down over the wires. A channel-grooved cover strip tacked to the tube protects the emergent leads. No. 22 B & S gage (0.025 in.), chromel-against-alumel, glass-insulated, duplex cable is used.

Sec. 8·3   PIERCED-IN JUNCTIONS   87

In Fig. 8·3,[9-11] a short hole is drilled of such a size as to admit a large, fused bead. The parent metal is then peened around this bead, embedding it tightly in place.

8·3   PIERCED-IN JUNCTIONS

Figure 8·4 [12] shows a thermocouple assembly intended for measuring operating temperatures in the rubber of automobile tires. A hole, in

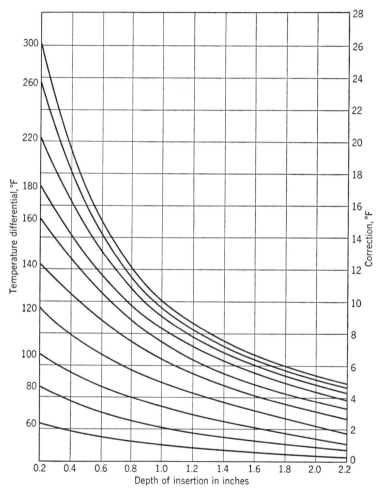

Fig. 8·5. Correction curves for automobile-tire thermocouple (see Fig. 8·4). *Temperature differential* is the difference between the junction temperature and that of the ambient air. (*By permission, Spear and Purdy in Ind. Eng. Chem.*, 15, *no. 8, 844, August, 1923.*)

which the ¼-in. diameter shank of the couple fits tightly, is drilled in the tire, replaced by a rubber plug, and the tire brought to operating temperature in road service. The plug is then replaced by the couple

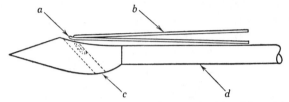

Fig. 8·6. Awl-inserted thermocouple. *a*, twisted and soldered junction; *b*, thermocouple wires; *c*, 1/32-in. hole drilled at 45° angle to permit couple to slip out of head when awl is withdrawn; *d*, head of awl.

and the temperature is determined by extrapolating back on the cooling curve, applying corrections per Fig. 8·5 [12] corresponding to depth of insertion. A 3/16-in. diameter, 1/64-in.-thick copper disk, to which the

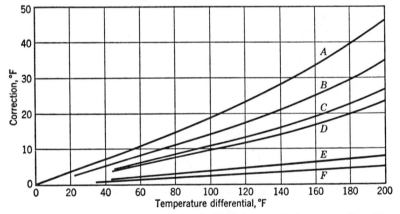

Fig. 8·7. Corrections for copper-constantan, awl-inserted thermocouples. *Temperature differential* is the difference between the junction temperature and that of the ambient air. *A*, No. 28 B & S gage (0.013 in.) inserted 0.30 in. *B*, No. 28 B & S gage (0.013 in.) inserted 0.55 in. *C*, No. 28 B & S gage (0.013 in.) inserted 0.85 in. *D*, No. 28 B & S gage (0.013 in.) inserted 1.15 in. *E*, No. 36 B & S gage (0.005 in.) inserted 0.55 in. *F*, No. 36 B & S gage (0.005 in.) inserted 1.15 in. (*By permission, Spear and Purdy in Ind. Eng. Chem., 15, no. 8, 844, August, 1923.*)

No. 36 or No. 40 B & S gage (0.005 in. or 0.003 in.) copper and constantan thermocouple wires are soldered, is intended to provide a maximum area of contact between the junction and the rubber. The shank is designed to insulate the leads thermally elsewhere than at the junction.

Figure 8·6 [13,14] shows arrangements for inserting a thermocouple with an awl, which pierces the necessary hole and on retraction leaves the thermocouple installed in position. Curves, Fig. 8·7,[12] indicate the magnitudes of error to be expected with thermocouples installed by this method in rubber of medium thermal conductivity.

The methods of Figs. 8·4, 6 are adapted generally to materials sufficiently elastic in character to press against the junction disk or wires after insertion, thus providing thermal contact. A similar result can often be achieved by pressing the wires between two layers of the soft material during its assembly.[15,16] The error curves, Figs. 8·5, 7, apply to a material, i.e., rubber, of relatively low thermal conductivity. An installation of the sort shown in Fig. 8·4, where a substantial bulk of low-conductivity material replaces the parent substance, would introduce an additional error due to this substitution, if made in good thermal conductors such as metallic bodies.

In biological work, where the installation is made in tissue, low thermal conductivities may also prevail. Electrical shunting through the semiconducting vital fluids must be considered. Very small size is often essential; for example, Saylor [17] installed junctions in microscope slides.

8·4 WELDED JUNCTIONS

Figure 8·8 [18] shows a method [19] used in the crowns of sodium-filled exhaust valves. The wires are led across the sodium cavity inside a stainless-steel tube, and welded into the surface of the valve crown.

Much unpublished work has been done on installations where the beads or ends of the thermocouple wires are electrically

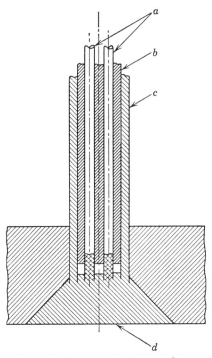

Fig. 8·8. Welded junction. *a*, thermocouple wires; *b*, two-hole ceramic tubing; *c*, stainless-steel tubing; *d*, welded junction.

welded directly to the parent metal. If the parent metal is one for which the thermocouple materials have a natural affinity, and if adequate equipment is used, good welds can be achieved. For example, the authors, in pulsation welding of copper wire to massive steel, have consistently produced welds stronger than the wire. Where a condenser is discharged through the wire by tapping the end or bead against the parent metal, welds tend to variable geometry and strength. If the parent metal is such that adhesion would not normally occur (when it is an aluminum alloy), the welds are, at best, unreliable. The massive electrodes, needed near to the weld in the pulsation technique, require much space; similarly, lateral electrical clearance is necessary in the relatively high-voltage condenser-discharge technique. Accessibility is required for rigorous inspection. These space requirements suggest limitation to surface applications. Many faulty data have resulted from broken welds in installations at the ends of drilled holes.

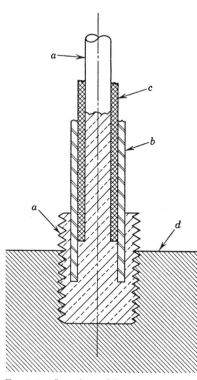

FIG. 8·9. Junction with parent metal as one lead. *a*, constantan wire; *b*, tubing of same material as parent body; *c*, ceramic tubing; *d*, parent body.

8·5 JUNCTIONS WITH PARENT METAL AS ONE LEAD

In the "one-wire" thermocouple (see Fig. 8·9),[20-22] the parent metal serves as the return lead.[23] To lessen errors resulting from the inhomogeneities in composition and temperature, which are inevitable in ordinary structures, an armoring tube of the parent-metal material short-circuits the structure (see Sec. 8·6 regarding the design of Akin and McAdams).[24] In a modification [25] of this method, the parent-metal tube is soldered or welded directly into a close-fitting hole in the parent metal; the joint includes the tip of the central constantan wire. In a similar design,[26] a tubular return lead may be insulated from the parent metal, except at the junction, and this need not necessarily be of the same material. Matched thermocouple materials are commercially available in this tube-enclosed wire form, incorporating a flexible insulating layer.[27] Although Gibson used $\frac{1}{4}$-in. diameter tubing,[26] later work has involved 0.083-in.,[22] 0.040-in.,[25] and smaller-sized tubing.

8·6 SOLDERED AND CEMENTED JUNCTIONS

Figure 8·10 [28] shows an arrangement where mercury is used to provide metallic contact between the junction and the parent metal. A

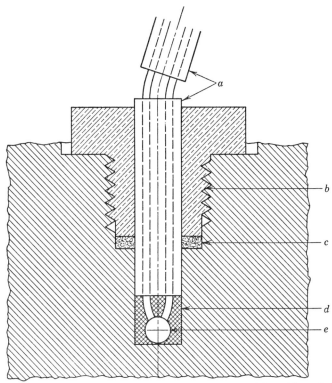

Fig. 8·10. Liquid-contact installation. *a*, two-hole porcelain tubing; *b*, hollow retaining screw; *c*, asbestos packing; *d*, mercury; *e*, junction.

stuffing box serves both to prevent leakage of mercury and to maintain mechanical integrity. The same design could be used with a solder or nonmetallic contact material operated above or below its melting point. Where the contact material failed to "wet" both junction and parent material, a thermal resistance would arise.[29] Adequate precautions are required to prevent loss of the mercury through the holes in the porcelain tubing. Mercury cannot be used with thermocouple materials with which it amalgamates, unless these are reliably protected by enamel.

Thermocouples laid in grooves milled in the surface of the parent material have been used for measuring tube-wall temperatures on heat exchangers. Under conditions of high surface transmission of heat, the leads for such installations may function as fins, where they emerge from the surface. If the point of such exit is near the location of the junction, the temperature at the point of measurement may thereby be altered. It is hence necessary that the point of emergence be sufficiently distant from the junction. It is likewise important that the conductivity immediately under the surface and the surface texture at the junction be altered as little as possible by the installation. Having the leads embedded in very nearly isothermal surfaces tends to minimize disturbance of the junction temperature by conduction of heat along the leads. Figures 8·11, 12, 13, 14 illustrate designs intended to meet these conditions.

In Fig. 8·11,[30, 31] showing a method developed by Baker and Mueller, the junction is soldered into a hole drilled tangentially from the terminus of a roughly helical groove. The brass tube protecting the leads is soldered in this groove and polished flush, emerging on the opposite side. The method of Patton and Feagan [32] differs in that the groove is longitudinal; thus, the leads emerge at points sufficiently distant from the junction so as not to disturb the fluid-flow pattern in that vicinity.

Akin and McAdams [24] sealed one wire into a Pyrex capillary exposed at the tip. This unit was then soldered into the hole so that the parent metal served as the return lead (see also Sec. 8·5, Fig. 8·9).

Mohun and Peterson, Fig. 8·12,[33] employed precision methods in locating the junction. A brass tube is used for centering. The junction is at a 0.02-in.-long web of solder between the wires. Lacquer coating separates the wires (including the junction) from the solder that fills the hole. The leads are embedded in cured resin in a longitudinal groove.

Hebbard and Badger, Fig. 8·13,[34, 35] soldered the junction in a chordal hole, admitting the solder through a radial perforation. The leads emerge on opposite sides. They are insulated, except at the junction, by a Bakelite-lacquered cotton-thread winding. They are then carried

Sec. 8·6　SOLDERED AND CEMENTED JUNCTIONS　93

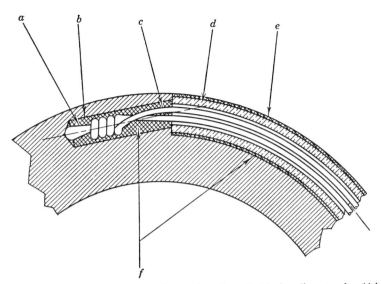

Fig. 8·11. Tube-wall thermocouple. *a*, junction; *b*, 1/16 in. diameter by 1/4-in.-deep drilled hole; *c*, thermocouple wires; *d*, 0.08 in. outside diameter, 0.05 in. inside diameter, brass tube; *e*, 3/32-in. by 3/32-in. groove; *f*, solder.

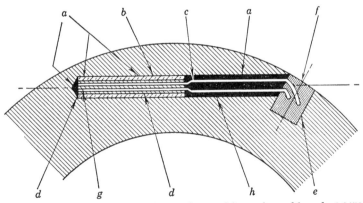

Fig. 8·12. Tube-wall thermocouple. *a*, low-melting-point solder; *b*, 0.0470 in. max, 0.0468 in. min, outside diameter, 0.0212 in. max, 0.0210 in. min inside diameter by 1/4-in. long brass tube; *c*, No. 30 B & S gage (0.010 in.), copper and constantan thermocouple wires; *d*, baked Heresite lacquer; *e*, cured resin; *f*, 1/16-in. by 3/32-in. groove; *g*, soldered junction; *h*, No. 56 (0.0465 in.) drill, 1/2 in. deep.

to the opposite side of the tube, embedded in Bakelite cement in an annular groove.

Figure 8·14 indicates an application of the method described in

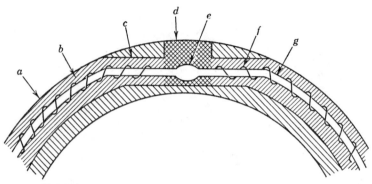

Fig. 8·13. Tube-wall thermocouple. *a*, 0.06-in.-wide by 0.08-in.-deep groove; *b*, Bakelite cement; *c*, No. 53 (0.060 in.) chord hole; *d*, lead-tin solder; *e*, butt-silver-soldered junction; *f*, cotton thread; *g*, No. 24 B & S gage (0.020 in.), copper and constantan thermocouple wires.

Sec. 11·8 (based on the experience of the authors). This modification is intended for tube walls where maximum precision is required under conditions of exposure to corrosive fluids, mechanical shock or vibra-

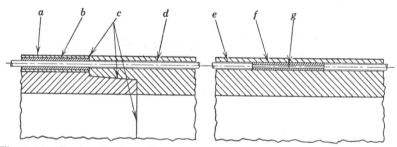

Fig. 8·14. Cemented tube-wall thermocouple. *a*, nickel tube silver-soldered in groove; *b*, "spaghetti," air-space, or cement; *c*, brazed joint; *d*, 0.016 in. outside diameter glass-fiber insulation over No. 30 B & S gage (0.010 in.), iron and constantan thermocouple wires; *e*, No. 76 (0.020) drilled hole; *f*, bare wire embedded in cement; *g*, butt-welded junction.

tion, and temperatures up to around 1400°F. For 0.010-in. diameter wire, the length of the longitudinally drilled insert section can be 1.5 in. If smaller wire and drill size are used, a proportionately shorter section will suffice. Similarly, on a larger scale, the section should be longer. After grooving and shouldering the two sections of the tube, the drilled insert is assembled by brazing. The butt-welded junction (see Sec. 5·8) is then threaded through and cemented in the hole. The protecting tubes are threaded on and silver-soldered in position, sealing in the wire. The space between the wire and the protecting tubes can be left free, or, for further protection, can be filled with "spaghetti" or cement.

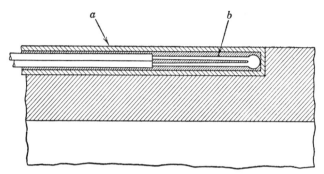

Fig. 8·15. Surface-groove thermocouple. a, $\frac{1}{16}$-in. outside diameter, $\frac{3}{64}$-in. inside diameter, nickel tube closed at end, brazed in $\frac{1}{16}$-in. by $\frac{1}{16}$-in. groove; b, thermocouple installation (see Sec. 11·2).

Reiher[36] soldered the junction in a small depression at the terminus of an 0.008-in. by 0.008-in. longitudinal groove. His No. 38 B & S gage (0.004-in.) copper and constantan thermocouple wires were laid in this groove embedded in cement. Colburn and Hougen[37,38] soldered the junction in a $\frac{1}{32}$-in. by $\frac{1}{32}$-in. annular groove with the leads emerging in opposite directions, embedded in glycerin-litharge cement. A number of other modifications in method have been used,[39] including one developed by Clement and Garland.[40]

Figure 8·15 (based on the experience of the authors) indicates a convenient and rugged design which has given satisfactory service in operation. This installation has been used in the outer surface of the tubular stems of sodium-filled exhaust valves. It does not satisfy the above stated requirements for precision under conditions of surface transfer of heat (see Sec. 8·6). Figure 8·16 presents a modification, equally rugged, nearly as convenient, and much more precise.

Figure 8·17 indicates a simple and commonly used method of installing a thermocouple in circumstances where soldering is feasible. For precision, it is necessary that the thermal current across the surface

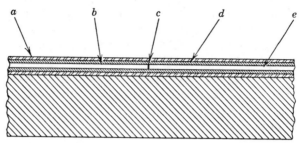

Fig. 8·16. Surface-groove thermocouple. *a*, $\frac{1}{64}$-in. outside diameter, 0.009-in. inside diameter, stainless-steel, hypodermic tubing silver-soldered in $\frac{1}{64}$-in. by $\frac{1}{64}$-in. groove; *b*, No. 38 B & S gage (0.004 in.) iron and constantan thermocouple wire; *c*, butt-welded junction; *d*, cement; *e*, silicone varnish on wire.

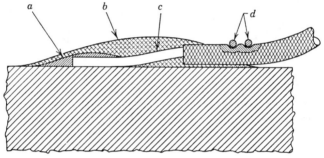

Fig. 8·17. Soldered thermocouple. *a*, solder or brazing material; *b*, G.E. 1201 Glyptal Red Enamel; *c*, thermocouple wires; *d*, wire binding.

be small in the neighborhood of the thermocouple and that the wire diameter be small in relation to the local massiveness of the parent metal.

8·7 PLUGS WITH JUNCTIONS

Figure 8·18 [41] illustrates one of several forms of a method [42-44] which involves soldering the wires into a metal fitting pressed against the parent material by a screw plug driven against it from the opposite direction.

Figures 8·19 [45] and 20 [46] show two forms of a method [47-49] that depends on soldering the junctions into a fitting. The fitting is subsequently screwed into the parent metal.

In the methods indicated by Figs. 8·21,[50] 22,[51] a screwed fitting serves to press the junction against the parent material.[52] By these

Sec. 8·7   PLUGS WITH JUNCTIONS   97

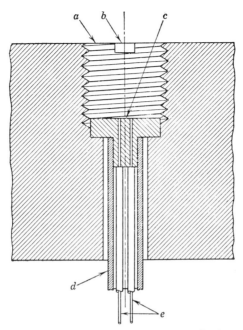

Fig. 8·18. Screw-pressure junction. *a*, pressure screw; *b*, slot; *c*, soldered junction; *d*, glass tubing; *e*, thermocouple wires.

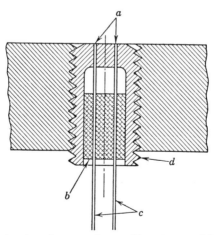

Fig. 8·19. Screw-plug junction. *a*, silver solder; *b*, porcelain; *c*, thermocouple wires; *d*, screw threads.

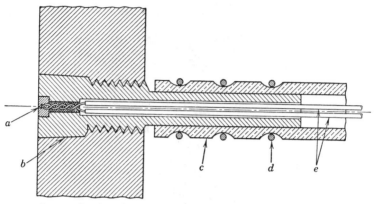

Fig. 8·20. Screw-plug junction. *a*, brazed junction; *b*, taper-screw plug; *c*, rubber tubing; *d*, binding wire; *e*, thermocouple wires.

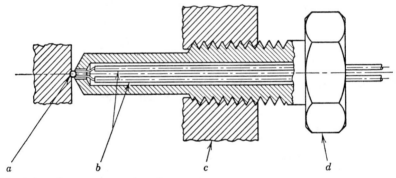

Fig. 8·21. Screw-pressure junction. *a*, beaded junction; *b*, thermocouple wires; *c*, yoke or parent body; *d*, screw.

arrangements, an installation can be made across a fluid jacket to indicate the temperature of the inner wall.[53]

8·8 TAPERED-PLUG JUNCTIONS

Figure 8·23 [54,55] indicates a thermocouple installed in a tapered plug made of material similar to the parent material.[56] The wires are individually riveted, the junction being formed through the plug material. It was hoped that the tapered joint would, by fitting, provide thermal contact [57] when tightly driven. Nägel [58] reports that such contact did not prove entirely adequate. According to Hug,[54] since the wires are in good thermal communication with the plug only at their riveted tips, they should lie for a "sufficient" distance in an isothermal plane.

Sec. 8·9 PRESSED-IN PLUG COUPLES

A geometrical temperature-pattern check was made on a 25-times-size-scale model.[54,55] No assurance exists, however, that thermal contact conditions in an installation would be the same as on the model.

Figure 8·24 [54] differs from the preceding figure in that the plug is in two parts with the junction squeezed between.[59]

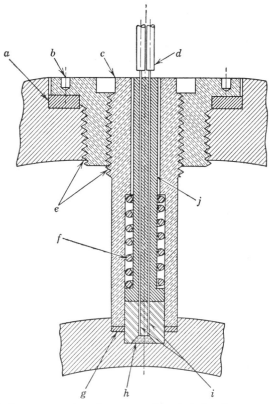

Fig. 8·22. Spring-pressure junction. *a*, gasket; *b*, holes for spanner wrench; *c*, hexagon for wrench; *d*, insulated thermocouple wires; *e*, screw bushings; *f*, pressure spring; *g*, gasket; *h*, junction brazed in plug; *i*, thermocouple wires; *j*, insulating plunger.

8·9 PRESSED-IN PLUG COUPLES

Figures 8·25,[60,61] 26 [60,61] indicate two variations of a design employing cylindrical plugs.[62-64] The material of the plug is the same as that in which the installation is made. In Fig. 8·25, the junction is brazed into the plug, the latter being pressed into the parent material. In Fig. 8·26 the wires are staked into separate plugs, which are in turn staked into the parent material. The junction occurs through a web in the parent material. Thermal contact, attempted only at the plugs, is of questionable adequacy under these circumstances.[57]

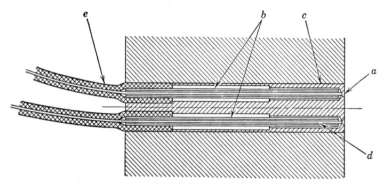

Fig. 8·23. Taper-plug thermocouple. *a*, riveted ends; *b*, quartz tubes; *c*, 1:50-taper plug; *d*, thermocouple wires; *e*, asbestos braid.

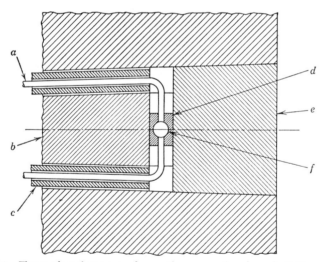

Fig. 8·24. Taper-plug thermocouple. *a*, thermocouple wires; *b*, 1:50-taper plug; *c*, quartz tubes; *d*, solder; *e*, 1:50-taper plug; *f*, junction.

8·10 LIMITATIONS OF METHODS

Although the authors of the designs described in this chapter have considered their methods effective for the particular measurements intended, none of these designs should be adopted for use without careful consideration of the requirements of the specific temperature

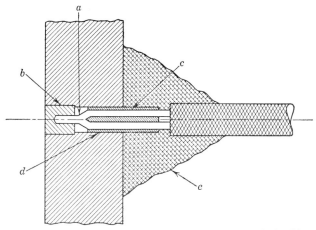

Fig. 8·25. Pressed-in-plug thermocouple. *a*, silver solder; *b*, bushing of same metal as parent body, pressed in place; *c*, thermocouple wires; *d*, Sillimanite tube; *e*, porcelain cement.

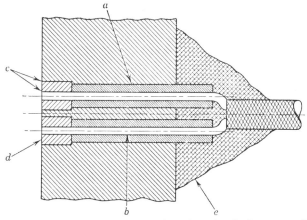

Fig. 8·26. Pressed-in-plug junction. *a*, Sillimanite tube; *b*, thermocouple wires; *c*, stake; *d*, bushing of same metal as parent body; *e*, porcelain cement.

measurement to be made. Each of these installations is subject to more or less severe limitations. Only rational analysis [65-72] and actual test [54, 66, 71, 73-75] can determine a suitable choice of design for any specific set of conditions (see Secs. 7·1 to 7·12; and 11·4).

REFERENCES

1. H. Kreisinger, and J. F. Barkley, "Heat Transmission through Boiler Tubes," *Technical Paper* 114, pp. 11–12, 25–28, U. S. Bureau of Mines, Washington (1915).
2. R. Royds, *The Measurement of Steady and Fluctuating Temperatures*, pp. 126–129, Constable & Co., London, 1921; summarizes Ref. 1.
3. Royds, pp. 110–114, 118–119 (see Ref. 2), summarizes work of Callendar (1904).
4. F. G. Shoemaker, "Method of Measuring Temperature of Pistons," *Army Air Forces Technical Report* 2825, p. 10, U. S. War Department, Air Corps Matériel Division, Dayton (June, 1927).
5. O. G. Kaasa, "Measurement of Metal Temperatures of Cracking Still Tubes," American Institute of Physics, *Temperature*, pp. 1091–1092, Reinhold Publishing Corp., New York, 1941.
6. R. M. Kleinberg, F. B. Kacena, and E. C. Lawson, Jr., "Technical Report on Heat Transfer to Liquid Metals," *Report* ME-T-1, pp. 7–9, U. S. Office of Naval Research Project NRO 35-323, Department of Mechanical Engineering, University of Delaware, Newark (1950).
7. C. G. R. Humphreys, "Thermocouples for Furnace-Tube Surface Temperature Measurements," *Combustion*, 16, no. 6, pp. 53–55 (December, 1944).
8. L. B. Schueler, "An Investigation of the Variation in Heat Absorption in a Pulverized-Coal-Fired Water-Cooled Steam-Boiler Furnace. I.–Variations in Heat Absorption as Shown by Measurement of Surface Temperature of Exposed Side of Furnace Tubes," *Trans. ASME*, American Society of Mechanical Engineers, 70, pp. 554–555 (July, 1948).
9. B. Pinkel and E. J. Manganiello, "A Method of Measuring Piston Temperatures," *Technical Note* 765, pp. 3–4, U. S. National Advisory Committee for Aeronautics, Washington (June, 1940).
10. E. J. Manganiello and E. Bernardo, "Cylinder Temperatures of Two Liquid-Cooled Aircraft Cylinders for Various Engine and Coolant Conditions," *Wartime Report* E-146, pp. 3–4, U. S. National Advisory Committee for Aeronautics, Washington (October, 1945).
11. M. F. Valerino and S. J. Kaufman, "Cylinder-Temperature and Cooling-Air-Pressure Instrumentation for Air-Cooled-Engine Cooling Investigations," *Technical Note* 1509, pp. 4–5, 29, U. S. National Advisory Committee for Aeronautics, Washington (January, 1948).
12. E. B. Spear and J. F. Purdy, "The Measurement of Temperature in Rubber and Insulating Materials by Means of Thermocouples," *Ind. Eng. Chem.*, 15, no. 8, p. 844 (August, 1923).
13. A. O. Ashman, "The Determination of Tire Temperatures," *Research Bulletin*, 10 pp., New Jersey Zinc Co., New York, undated.
14. A. O. Ashman, "A Method of Measuring the Temperature at Different Points in the Body of an Automobile Tire," *Research Bulletin*, pp. 3–5, 11–14, New Jersey Zinc Co., New York, November, 1921; published in *Rubber Age*, 10, pp. 161–163 (December, 1921).
15. M. E. Dunlap and E. R. Bell, "Temperature Distribution in White-Oak Laminated Timbers Heated in a High-Frequency Field," *Paper* 46-A-68, *Trans. ASME*, American Society of Mechanical Engineers, 69, no. 5, p. 510 (July, 1947).

REFERENCES

16. A. M. Stoll and J. D. Hardy, "Direct Experimental Comparison of Several Surface Temperature Measuring Devices," *Rev. Sci. Instr.*, 20, no. 9, pp. 680–681 (September, 1949).
17. C. P. Saylor, "Control and Measurement of Temperature Under the Microscope," American Institute of Physics, *Temperature*, pp. 675–676, 678, Reinhold Publishing Corp., New York, 1941.
18. J. C. Sanders, H. D. Wilsted, and B. A. Mulcahy, "Operating Temperatures of a Sodium-Cooled Exhaust Valve as Measured by a Thermocouple," *Wartime Report* E-140, pp. 4–5, U. S. National Advisory Committee for Aeronautics, Washington (December, 1943).
19. N. MacCoull, "Power Loss Accompanying Detonation," *S.A.E. Journal*, Society of Automotive Engineers, 44, no. 4, p. 160 (April, 1939).
20. B. A. Mulcahy and M. A. Zipkin, "Tests of Improvements in Exhaust-Valve Performance Resulting from Changes in Exhaust-Valve and Port Design," *Wartime Report* E-45, p. 2, Fig. 1, U. S. National Advisory Committee for Aeronautics, Washington (September, 1945).
21. M. A. Zipkin and J. C. Sanders, "Correlation of Exhaust-Valve Temperatures with Engine Operating Conditions and Valve Design," *Wartime Report* E-48, Fig. 3, U. S. National Advisory Committee for Aeronautics, Washington (October, 1945).
22. M. D. Peters, "Effect of Increasing the Size of the Valve-Guide Boss on the Exhaust-Valve Temperature and the Volumetric Efficiency of an Aircraft Cylinder," *Wartime Report* E-61, p. 5, U. S. National Advisory Committee for Aeronautics, Washington (February, 1945).
23. Royds, pp. 115–118 (see Ref. 2), summarizes work of Hopkinson.
24. G. A. Akin and W. H. McAdams, "Boiling: Heat Transfer in Natural Convection Evaporator," *Trans. Am. Inst. Chem. Engrs.*, 35, pp. 138–140 (1939).
25. A. T. Sutor, L. C. Corrington, and C. Dudugjian, "Exhaust-Valve Temperatures in a Liquid-Cooled Aircraft-Engine Cylinder as Affected by Engine Operating Variables," *Technical Note* 1209, pp. 3–4, U. S. National Advisory Committee for Aeronautics, Washington (February, 1947).
26. A. H. Gibson and H. W. Baker, "Exhaust-Valve and Cylinder-Head Temperatures in High-Speed Petrol Engines," *Inst. Mech. Engrs. Proc.*, II, pp. 1052–1055 (December, 1923).
27. Leeds & Northrup Co., "Thermocouples—Assemblies Parts and Accessories," Catalog no. EN-S2, pp. 13, 19, 30, Philadelphia (1952).
28. L. A. Wendt and T. B. Rendel, "Some Notes on Piston Temperature and Its Measurement," Preprint, p. 2, *Piston Temperature Symposium at Annual Meeting of Society of Automotive Engineers*, Detroit, January, 1939.
29. N. D. Weills and E. A. Ryder, "Thermal Resistance Measurements of Joints Formed between Stationary Metal Surfaces," *Paper* 48-SA-43, *Trans. ASME*, American Society of Mechanical Engineers, 71, no. 3, p. 260 (April, 1949).
30. E. M. Baker and A. C. Mueller, "Condensation of Vapors on a Horizontal Tube—Heat Transfer Coefficients for the Condensation of Mixed Vapors of Immiscible Liquids," *Ind. Eng. Chem.*, 29, no. 9, pp. 1067–1068 (September, 1937).
31. E. M. Baker and A. C. Mueller, "Condensation of Vapors on a Horizontal Tube. Part II: Heat Transfer Coefficients for the Condensation of Mixed

Vapors of Immiscible Liquids," *Trans. Am. Inst. Chem. Engrs.*, 33, pp. 542–543 (1937).
32. E. L. Patton and R. A. Feagan, Jr., "A Method of Installing Tube-Wall Thermocouples," *Ind. Eng. Chem., Anal. Ed.*, 13, no. 11, pp. 823–824 (November, 1941).
33. W. A. Mohun and W. S. Peterson, "Precision of Heat Transfer Measurements with Thermocouples—Geometric Errors," *Can. Chem. Process Ind.*, 31, no. 10, pp. 908–913 (October, 1947).
34. G. M. Hebbard and W. L. Badger, "Measurement of Tube Wall Temperatures in Heat Transfer Experiments," *Ind. Eng. Chem., Anal. Ed.*, 5, no. 6, pp. 359–361 (November, 1933).
35. G. M. Hebbard and W. L. Badger, "Steam Film Heat Transfer Coefficients for Vertical Tubes," *Ind. Eng. Chem.*, 26, no. 4, pp. 421–422 (April, 1934); also published in *Trans. Am. Inst. Chem. Engrs.*, 30, pp. 198–199 (1933–1934).
36. H. Reiher, "Wärmeübergang von Strömender Luft an Rohre und Röhrenbündel im Kreuzstrom" (Heat Transfer from Flowing Air to Tubes and Tubebanks in Crossflow), *Forschungsarbeiten auf dem Gebiete des Ingenieurwesens*, 269, pp. 15–16 (1925).
37. A. P. Colburn and O. A. Hougen, "Studies in Heat Transmission Particularly as Applied to Tubular Gas Condensers," *Engineering Experiment Station Series Bulletin* 70, pp. 69–70, University of Wisconsin, Madison (October, 1930).
38. A. P. Colburn and O. A. Hougen, "Studies in Heat Transmission. I.—Measurement of Fluid and Surface Temperatures," *Ind. Eng. Chem.*, 22, no. 5, pp. 522–524 (May, 1930). This article is a condensed version of Ref. 37.
39. S. C. Hyman and C. F. Bonilla, "Heat Transfer by Natural Convection from Horizontal Cylinders to Liquid Metals; Final Report for July 1, 1949, to June 30, 1950," *Report* NYO-560, pp. 27–41, U. S. Atomic Energy Commission, Technical Information Service, Oak Ridge (June, 1950).
40. J. K. Clement and C. M. Garland, "A Study in Heat Transmission—The Transmission of Heat to Water in Tubes as Affected by the Velocity of the Water," *Engineering Experiment Station Bulletin* 40, University of Illinois, Urbana (September, 1909). Summarized in Royds, p. 122 (see Ref. 2).
41. W. Riehm, "Temperaturmessungen an Kolben von Oelmaschinen" (Temperature Measurements in Pistons of Oil Engines), *Z. Ver. deut. Ing.*, 65, no. 35, p. 923 (August, 1921).
42. F. R. B. Watson, "Cylinder Temperatures in a 25 B.H.P. Crude-Oil Engine, and Their Effect on Engine Performance," *Inst. Mech. Engrs. Proc.*, II, p. 938 (December, 1928).
43. H. W. Baker, "The Operating Temperatures of Cast Iron and Aluminum Pistons in a 12-Inch Bore Oil Engine," *Inst. Mech. Engrs. Proc.*, 127, pp. 222–230 (November, 1934).
44. P. V. Keyser and E. F. Miller, "Piston and Piston-Ring Temperatures," *J. Inst. Petroleum*, 25, no. 194, p. 781 (December, 1939).
45. E. L. Bass, "Fuels for Aircraft Engines," *J. Roy. Aeronaut. Soc.*, 39, pp. 934–935 (1935).
46. A. W. Judge, *The Testing of High Speed Internal Combustion Engines*, 3d edition, pp. 345–347, Chapman & Hall, London, 1943; describes Gibson's method.

REFERENCES

47. G. F. Mucklow, "Piston Temperatures in a Solid-Injection Oil-Engine," *Inst. Mech. Engrs. Proc.*, 123, pp. 353–357 (December, 1932).
48. H. W. Baker, "Piston Temperatures in a Sleeve Valve Oil Engine," *Inst. Mech. Engrs. Proc.*, 135, pp. 39–45 (January, 1937).
49. W. L. Bride, "Piston Crown Temperatures in a Compression-ignition Engine with 'Comet' Head," *Inst. Mech. Engrs. J. & Proc.*, 150, no. 4, p. 135 (*J:* February, 1944, *Proc.:* 1943).
50. Judge, pp. 346–347 (see Ref. 46).
51. Keyser and Miller, p. 782 (see Ref. 44).
52. Valerino and Kaufman, pp. 3–4, 19 (see Ref. 11).
53. Royds, pp. 120–122 (see Ref. 2), summarizes work of Jordan (1909).
54. K. Hug, *Messung und Berechnung von Kolbentemperaturen in Dieselmotoren* (Measurement and Calculation of Piston Temperatures in Diesel Motors), Dissertation of the Eidgenossische Technische Hochschule Zurich, pp. 13–19, A. G. Gebr. Leeman & Co., Zurich and Leipzig, 1937.
55. G. Eichelberg, "Investigations on Internal-Combustion Engines—Some New Investigations on Old Combustion-Engine Problems—I," *Engineering*, 148, no. 3850, pp. 463–464 (Oct. 27, 1939).
56. M. O. Teetor, "Cylinder Temperature," *S.A.E. Trans.*, Society of Automotive Engineers, 39, no. 2, pp. 328–332 (August, 1936).
57. Weills and Ryder, pp. 259–267 (see Ref. 29).
58. A. Nägel, "The Transfer of Heat in Reciprocating Engines—III," *Engineering*, 127, no. 3294, p. 280 (Mar. 1, 1929).
59. E. M. Bryant, "On the Thermal Conditions of Iron, Steel and Copper when acting as Boiler-Plate," *Proc. Inst. Civ. Engrs.*, 132, p. 276 (1898). Summarized in Royds, p. 125 (see Ref. 2).
60. A. F. Underwood and A. A. Catlin, "Instrument for the Continuous Measurement of Piston Temperatures," *S.A.E. Trans.*, Society of Automotive Engineers, 48, no. 1, pp. 21–22 (January, 1941).
61. "Measuring Piston Temperatures," *Mech. Eng.*, 63, no. 3, pp. 219–220 (March, 1941); summarizes Underwood and Catlin (see Ref. 60).
62. Valerino and Kaufman, pp. 2, 15 (see Ref. 11).
63. J. H. Povolny and L. J. Chelko, "Cylinder-Head Temperatures and Coolant Heat Rejections of a Multicylinder, Liquid-Cooled Engines of 1710-Cubic-Inch Displacement," *Technical Note* 1606, pp. 4–5, 28–29, U. S. National Advisory Committee for Aeronautics, Washington (June, 1948).
64. J. H. Povolny, L. J. Bogdan, and L. J. Chelko, "Cylinder-Head Temperatures and Coolant Heat Rejection of a Multicylinder Liquid-Cooled Engine of 1650-Cubic-Inch Displacement," *Technical Note* 2069, p. 6, U. S. National Advisory Committee for Aeronautics, Washington (April, 1950).
65. I. B. Smith, "Applications and Limitations of Thermocouples for Measuring Temperatures," *Trans. Am. Inst. Elec. Engrs.*, 42, pp. 352–353 (February, 1923).
66. N. P. Bailey, "The Response of Thermocouples," *Mech. Eng.*, 53, no. 11, pp. 797–804 (November, 1931).
67. M. Jakob, "The Influence of the Free End of a Rod on Heat Transfer," *Heat Transfer: Research Publications, Illinois Institute of Technology*, 2, pp. 230–237 (1942).

68. Mohun and Peterson, pp. 911–913 (see Ref. 33).
69. L. M. K. Boelter, F. E. Romie, A. G. Guibert, and M. A. Miller, "An Investigation of Aircraft Heaters, XXVIII—Equations for Steady-State Temperature Distribution Caused by Thermal Sources in Flat Plates Applied to Calculation of Thermocouple Errors, Heat-Meter Corrections, and Heat Transfer by Pin-Fin Plates," *Technical Note* 1452, pp. 9–15, U. S. National Advisory Committee for Aeronautics, Washington (August, 1948).
70. W. A. Mohun, "Precision of Heat Transfer Measurements with Thermocouples—Insulation Error," *Can. J. Research*, Sec. F, 26, Sec. F, pp. 565–583 (December, 1948).
71. H. D. Baker and E. A. Ryder, "A Method of Measuring Local Internal Temperatures in Solids," *Paper* 50-A-101, pp. 1–2, 5–8, American Society of Mechanical Engineers, New York (1950).
72. F. Lieneweg, "Die Bestimmung von Temperaturmessfehlern mittels Thermometer-Kennzahlen" (The Determination of Errors in Temperature Measurement in Terms of Thermometer Parameters), 28 pp., Sonderdruck aus der Festschrift W. C. Heraeus CMBH., Hanau aus, Anlass ihres hundert jährigen Bestehens am 1. 4. 1951.
73. Spear and Purdy, pp. 843–844 (see Ref. 12).
74. Ashman, Appendix (see Ref. 13).
75. L. E. Smith, "Heat-Conduction Errors in Temperature Measurements," *Paper* 49-S-35, *Trans. ASME*, American Society of Mechanical Engineers, 72, no. 1, pp. 71–76 (January, 1950).

9

DRILLING TECHNIQUE

9·1 INTRODUCTION

Insertion of the sensitive element for temperature measurement at an interior point in a solid body usually requires that a hole be made in the body. Thus, to install an element in an engine cylinder, turbine blade, or die mold, it is necessary to drill a hole to the interior point. There are, however, exceptional situations where it is not necessary that a hole be drilled. If the material composing the body is soft or granular, the sensitive element may be able to "spear" its own hole. If the body is composite, it may be possible to assemble the body about the sensitive element and the leads, clamping these between mating surfaces either as sheets or in grooves or recesses cut in the surfaces. Sometimes it may be possible to cast part or all of the body about the element and leads.

If a hole is drilled, the required depth of the hole is the distance, from the nearest accessible point on the surface from which the leads are to emerge, to the interior point at which temperature is to be measured. The diameter of the hole must be sufficient to accommodate the sensitive element and the leads.

The larger the diameter of the hole, the greater will be the disturbance in temperature distribution within the body, because of the existence of the hole and of the thermocouple installation. Error in measurement of the location at which a temperature occurs is thereby greater for a larger-diameter hole, and less for a smaller-diameter hole. The required diameter of hole is that which will just accommodate the installation. With thermocouples, the size of the sensitive element and leads, i.e., of the bead and lead wires, can be made very small by using small-diameter wires. Satisfactory means are available for installing thermocouples in holes only slightly larger than sufficient to accommodate the bead and the lead wires.

9·2 REQUIREMENTS

To achieve the required degree of precision in measurement of the location at which the indicated temperature exists, it is necessary to be able to measure with a corresponding degree of precision the location of the bottom of the drilled hole where the bead or other sensitive element is placed. This implies measurement of (1), the location of

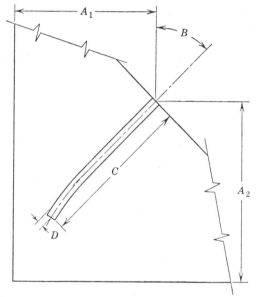

Fig. 9·1. Location of the bottom of a drilled hole.

the mouth of the hole; (2), the direction of the tangent to the axis of the hole at the mouth; (3), the depth of the hole; and, (4), the distance that the bottom of the hole is displaced from the axis (see A_1 and A_2, B, C, and D, respectively, Fig. 9·1).

The displacement of the bottom of the hole, D, may result from *drift* or *runout* in the drilling, i.e., failure to achieve a straight hole. Usually, it is permissible to measure temperature at a point slightly displaced from the intended point, provided that one knows where the bottom of the hole finally turns out to be. In deep, small-diameter holes, however, the quantity D is difficult to measure quantitatively. Hence, for such holes it is ordinarily best to use drilling technique which can be depended upon to yield sufficiently straight holes, and let D stand as a limited error after the degree of straightness has been checked by suitable means.

Thus, in general, the drilling technique must be such as to meet the following six requirements: (1) holes must be drilled small in diameter, i.e., the diameter which will just accommodate the thermocouple installation; (2) holes must be drilled very straight, with a large ratio of depth to diameter; (3) holes must be accurately located for dimensions A and B (see Fig. 9·1) in bodies in which it is desired to measure temperature, despite the diversity of the shapes and sizes of the bodies; (4) holes must be drilled in the various types of materials in which it may be desired to measure temperature; (5) chips and cutting fluid must be removable from the finished holes; and (6) it must be possible to make accurate final determination of the dimensions A_1 and A_2, B, C, and D (see Fig. 9·1) for the finished hole. Means for satisfying these six requirements will be discussed in the succeeding sections of this chapter.

9·3 DEEP HOLES

The drilling of very deep holes several inches in diameter has become a highly developed art in the oil fields. Drilling holes a few thousandths of an inch in diameter and a few hundredths of an inch deep is necessary for making rayon-fiber spinnerets and Diesel-engine jets.[1] Drilling holes with great precision a few tenths of an inch in diameter and several feet deep is required in the manufacture of rifle barrels.[2-4] Drilling holes a few hundredths of an inch in diameter and several inches deep is often necessary in thermocouple work.[5]

Success in drilling deep, small-diameter holes depends on three factors: (1), the drill used; (2), the drilling machine; and, (3), the skill of the operator.

The methods to be described first apply to drilling in metals such as brass, bronze, copper, aluminum, iron, tool and alloy steels in the annealed condition, wood, fiber, plastics, soapstone, etc. When very hard or abrasive materials are to be drilled, special methods must be employed.

9·4 TYPES OF DRILLS

American-made carbon steel and high-speed (HS) steel *twist drills* (see Fig. 9·2) are available in standard sizes ranging from several inches down to 0.013 in. in diameter. Swiss-made steel drills can be obtained with diameters as small as 0.006 in. *Pivot-point* drills (sometimes called chisel-point, diamond-point, flat, or fishtail drills, see Fig. 9·3) are available commercially in sizes from 0.021 down to 0.001 in. diameter in steps of 0.0001 in.[6]

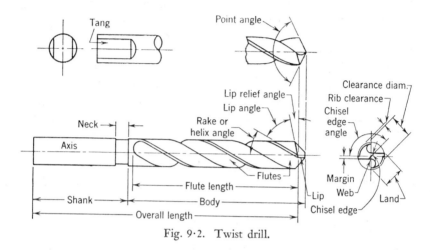

Fig. 9·2. Twist drill.

Standard twist drills (less than 3 in. diameter) are made nearly uniform in diameter for the entire length, the shank being relieved a very

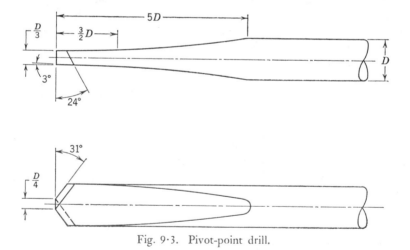

Fig. 9·3. Pivot-point drill.

small amount to prevent "binding." They come in standard lengths for the various sizes. Thus, the No. 80 (0.013 in.) drill comes 0.75 in. long, the No. 60 (0.040 in.) drill comes 1.5 in. long, and No. 1 (0.228

in.) drill comes 4.5 in. long. Drills in the various sizes may, on special order, be furnished in any desired length.

Small commercial pivot-point drills are made with enlarged shanks, the cutting blade being only a few hundredths of an inch long, i.e., about 7 diameters. Their use is accordingly limited to the drilling of very shallow holes.

9·5 MAKING SPECIAL DRILLS

It is possible to make pivot-point drills in any length (see Fig. 9·3) by hand grinding from standard drill rod in the following manner: (1) cut the desired length from a piece of drill rod of the correct diameter. Do this with a small three-cornered file in order to avoid bending the rod; (2) harden the tip of this piece of rod. The heat treatment used for this purpose should be that specified for the actual composition of the high-speed (HS) steel rod used. This usually involves approximately the succeeding operations; (3) preheat slowly to 1450 or 1650°F (light red); (4) raise quickly to 2250 or 2350°F (white) and hold at this in this temperature range for a fraction of a minute; (5) quench the tip of the rod in a molten bath at 1100°F (dark red); (6) cool slowly to 200 or 300°F; (7) slowly reheat for tempering to 1025 or 1150°F (dark red) and allow to cool; (8) to secure the requisite pivot-point shape at the lower end of the drill, grind on an ordinary tool grinder. As the thickness of the tip approaches the desired size (see Fig. 9·3), measure it frequently with a micrometer. The tapered length can be measured with a common scale. When this procedure is employed, symmetry, although important, must be judged by eye; (9) sharpen the tip in the same way as for a standard twist drill (see Sec. 9·6). Thus, pivot-point drills of any desired diameter and length can be made in a shop, without special equipment.

Hardening can also be accomplished by heating to the maximum temperature possible for small rod in an ordinary air-gas Bunsen flame, quenching in water, and then tempering as in (7) above. The use of this crude method makes it very difficult, however, to achieve anything approaching maximum strength and toughness in the cutting blade. As a result, a fragment of the cutting lip may break off and remain embedded in the bottom of the hole.

It is also possible to make extra-long, i.e., *extended shank*, drills by silver-soldering an extension onto the shank of a standard twist drill (see Fig. 9·4). The extension should be made from drill rod about 95 per cent of the drill diameter, thus making the diameter of the extension 5 per cent smaller than that of the drill, i.e., about 0.001 or 0.002 in. less for small drills.

A hole should be drilled in the end of this rod with a diameter about three-fourths the diameter of the drill, and two diameters deep. The end of the drill shank should be turned down to a close fit in this hole. Both of these operations are best performed on a precision bench lathe. Extreme care should be taken to achieve concentricity.

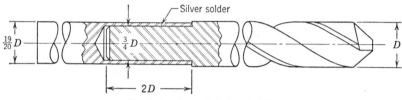

Fig. 9·4. Extended-shank drill.

For silver-soldering the joint, use flux No. 43 manufactured by Krembs & Company (Chicago, Illinois) or other suitable flux (see Sec. 5·8). The mating ends should be cleaned in carbon tetrachloride, painted with flux, heated to dull redness to melt the flux, then touched with fine silver-solder wire. This provides a thin coating of the silver solder. Chuck the two pieces in the headstock and tailstock, respectively, of a small bench lathe. Heat the mating ends with a small flame to melt the silver-solder coating. Then assemble quickly, pressing the parts together by advancing the handwheel-operated screw in the tailstock.

9·6 DRILL SHARPENING

The most precise drill sharpening is done by hand by an expert. In the absence of an expert, however, a reasonably satisfactory job is obtained more quickly and easily by machine. A typical drill-sharpening machine is shown in Fig. 9·5. This type of machine sharpens drills in sizes ranging from $11/32$ in. down to No. 72 (0.025 in.).

Standard twist drills in the smaller sizes are sufficiently cheap so that, when a drill-sharpening machine is not available, it is usually more economical to replace the drills than to resharpen them by hand after they become dulled. On the other hand, the relatively expensive special-length drills should not be discarded, and must be resharpened. The hand-made pivot-point drills must also be sharpened and resharpened. Both of these may be required in diameters smaller than can be sharpened by machine.

Sec. 9·6   DRILL SHARPENING   113

Drills of 0.040 in. diameter and larger can be sharpened on a small, fine-grained power grindstone by the following procedure. With the drill conveniently mounted in a pin vise, use the same motion as for a large drill. Hold the cutting lip (see Fig. 9·2) against the stone parallel to the axis of the grinder shaft with the pin-vise handle depressed sufficiently to provide the proper lip-relief angle. For the

Fig. 9·5. Drill-sharpening machine. (*Courtesy of Edward Blake Co., West Newton, Mass.*)

smaller-sized drills, use very light pressure and brief touches against the wheel. Inspect the work frequently, using a good pocket microscope.

For drills smaller than No. 60 (0.040 in.) the power grinder cuts too rapidly. It is also difficult to observe the small cutting lips while grinding. Therefore, instead of the power grinder, use a flat, hard Arkansas stone with light cutting oil such as kerosene. The pin vise carrying the drill should be held in the right hand exactly like a lead pencil and moved over the stone as in writing. Looking through a good pocket microscope, orient the cutting lip at the start of the operation. Grind the lips alternately in small installments by counting strokes across the stone and rotating the vise 180° in the hand. Inspect the work with a pocket microscope.

On small drills, the drill-point angle (instead of being 118° as for large drills) can vary from 90° to 135°, depending on the material. The lip-relief angle should be from 5° to 15°. These small drills should always be made of high-speed (HS) steel.

9·7 USING SMALL DRILLS

Small-diameter drills, regardless of their length, are very fragile as to torque resistance. If they are at all long, i.e., over 20 diameters, their flexibility disposes them to buckle except as supported by the metal being drilled. When the drill is cutting, the unsupported length between the chuck and the hole should not exceed 5 drill diameters. If the required hole is enlarged at the mouth, it should be drilled to the minimum diameter first and then counterbored.

The effects of excess pressure, i.e., excess force pushing the drill into the work, are (1), to increase the cut, thereby increasing the torque and tending to break the drill; (2), to cause the drill to drift or run out producing a crooked hole; and (3), to cause bellmouthing at the entrance to the hole. Thus, excess pressure may result in a crooked, bellmouthed hole though the drill does not break. The smaller the drill, the lighter the permissible pressure.

Even when the drill is perfectly sharp, the lightness of the pressure that can be applied with small drills results in removal of only a very small amount of metal per revolution. At these necessarily light pressures, an even slightly dull drill cuts substantially more slowly than a sharp drill. For lack of any convenient means for judging drill pressure, the operator tends to apply sufficient pressure to make the drill "cut." Thus, even a very skillful operator will tend to apply more pressure on a slightly dulled drill than on a perfectly sharp one. Since a drill which is not perfectly sharp therefore becomes a hazard, it is essential that the drill be kept entirely sharp at all times.

Inhomogeneities in the material being drilled, i.e., "hard spots" or "holes," will tend to deflect a drill (particularly a small drill), resulting in a crooked hole and possibly a broken drill. It is best not to attempt to drill in inhomogeneous material. If it is necessary to attempt to do so, however, the chances for success will be greatly increased by maximum drill speed, minimum drill pressure, and a perfectly sharp drill.

The cutting lips of a small drill (see Fig. 9·2), operated under light pressure, detach metal somewhat as do the individual particles of abrasive in a grinding wheel. Chips take the form of a fine powder which tends to accumulate in the flutes of a twist drill (see Fig. 9·2) or in the space on either side of the blade in a pivot-point drill (see Fig. 9·3). Unless removed, this powder packs tighter and tighter within this space. It presses outward against the walls of the hole until the friction thus developed becomes sufficient to prevent rotation of the drill. At this point the drill will fail in torsion, leaving the end broken off at the bottom of the hole.

The helical flutes (see Fig. 9·2) of a twist drill tend to "pump" this powder out of the hole. It is possible for them to do this only so long as the end of the fluted portion projects from the mouth of the hole. When the flutes are entirely filled with chips, the torque required to produce this pumping action, added to the torque required for cutting, may be too great for the shank of a small drill to withstand.

Since pivot-point drills do not have helical flutes, the pumping action does not occur as with the twist drills. Instead, the chip powder flows upward into the space on either side of the blade as the newly cut powder forces it up from below. Before the space is entirely filled, the pressure of this powder against the sides of the hole may develop sufficient friction to twist off the end of the drill. It may also very much reduce the rate of cutting.

In drilling deep, small-diameter holes, it is therefore necessary to remove chips by withdrawing the drill. This should be done frequently enough to prevent pressure, and thereby friction, from developing in the space filled by the chips. As often as sufficient material has been detached by the cutting lips to fill one-fourth to one-third of the chip space, the drill should be withdrawn and cleaned. If the drill is sharp and is cutting properly, withdrawal should occur after every few hundred revolutions of the spindle. At suitable operating speeds for small drills (see Sec. 9·8), this will mean that the drill should be withdrawn after every 1 or 2 seconds in action. Such a periodic up-and-down motion can be performed automatically (see step-drilling attachment, G and H, on Fig. 9·6).

A pipe cleaner of the type intended for cleaning the duct of a tobacco pipe can be used conveniently to remove chips from either twist or pivot-point drills, and simultaneously to supply the cutting fluid. After dipping in the cutting fluid, the pipe cleaner need merely be touched lightly against the tip or fluted portion of the rotating drill. Although for this purpose it is usually preferable to remove a twist drill entirely from the hole, the chips can be removed from a twist drill if it is raised sufficiently so that part of the fluted portion emerges and is touched with the pipe cleaner. A pivot-point drill, however, should always be withdrawn completely, because the chips tend to collect near the tip.

A cutting fluid of low viscosity should be used, the proper variety being determined by experiment. A furnace oil of about 36 or 40

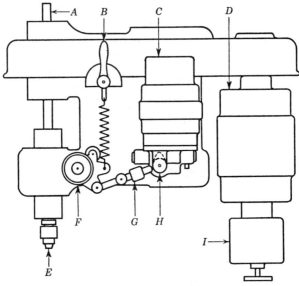

Fig. 9·6. Drill-head assembly. *A*, drill spindle; *B*, spring-tension handle; *C*, automatic backing-out motor; *D*, motor; *E*, drill chuck; *F*, manual feed; *G*, step-drilling-attachment backing-out follower; *H*, step-drilling-attachment backing-out cam; *I*, governor.

gravity will prove adequate for drilling in many types of materials. Kerosene is ordinarily quite satisfactory. Four parts turpentine to 1 part soluble oil has been used successfully in drilling tough die steel.

9·8 THE DRILLING MACHINE

A drilling machine for use in thermocouple work must meet definite specifications. The operator must be provided with the most sensitive possible means for judging drill pressure. The spindle must move up and down easily and be arranged so as to allow delicate fingertip control. A lamp must also be provided so that the operator has the best possible vision of the mouth of the hole and the exposed portion of the drill. For reasonable speed in drilling, it must be feasible with one quick, easy motion of the fingers to withdraw the drill its entire length for cleaning.

Sec. 9·8 THE DRILLING MACHINE 117

Since the drill pressure must necessarily be light and the cutting rate consequently very slow, it is essential that the drilling machine be so constructed as to permit drilling as nearly as possible at the full permissible cutting speed for the material being drilled. Even at high cutting speeds the drilling of deep small-diameter holes requires much patience on the part of the operator. Thus, drilling at the rate of an inch per hour is commonplace.

Because of the short radius on a small drill from the outer end of the cutting lip to the axis of rotation, the permissible cutting speed for high-speed (HS) steel can only be achieved by a relatively high speed of rotation. The drilling machine must thus be able to reach 40,000 or 50,000 rpm. On the other hand, for drilling holes as large as No. 54 (0.055 in.) in certain very tough alloy steels, the highest permissible speed may be as low as 700 rpm. Hence, the machine must also permit variation of the speed down to 700 rpm, and by small steps.

A long drill seems rarely to remain perfectly straight for more than a short time in service. While running freely in air at a suitably high speed almost any drill becomes perfectly straight, because of dynamic effects. This phenomenon may contribute slightly to the production of straight holes. Nevertheless, there is always a *critical speed* at which an unsupported drill vibrates laterally like a reed. If this vibration is permitted to develop unchecked, it will often cause the drill to swing out horizontally at the point of emergence from the chuck (see E on Fig. 9·6). The more slender the drill and the longer the exposed portion, the lower is the critical speed. If the exposed portion of the drill is supported at both ends, that is, by the chuck and the hole, the critical speed will be much higher than when the drill is supported only at the chuck end.

In order to raise the critical speed as high as possible and also to minimize the flexing of the drill with the consequent tendency to drift or run out, the drill should project from the chuck by an amount only slightly greater than the depth of the hole. Thus, when the drill is cutting, the unsupported length between the chuck and the entrance to the hole should not exceed 5 drill diameters. As the depth of the hole increases, the drill should be made to project more and more from the chuck to correspond to the increased depth of the hole. The resulting successive increases in the length of the drill may require corresponding reductions in speed to avoid critical speeds, as described in the preceding paragraph.

Any misalignment between the chuck and the axis of the hole must be taken up by bending of the drill in this short unsupported length, back and forth, every revolution of the spindle (see *A* on Fig. 9·6). If misalignment is appreciable, the drill is likely soon to break off at the entrance to the hole; it is difficult to extract the remains of such a broken drill from the hole.

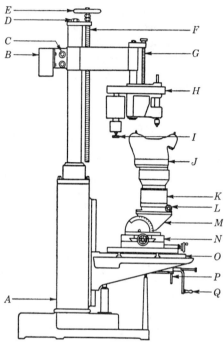

Fig. 9·7. Drilling machine. *A*, base; *B*, counterweight; *C*, clamping screws; *D*, clamping screw (permits 180° swing); *E*, handwheel; *F*, jackscrew; *G*, vertical adjusting screw; *H*, drill-head assembly; *I*, speed adjustment; *J*, work piece; *K*, holding fixture; *L*, rotary table; *M*, adjustable angle-tilting table; *N*, compound table; *O*, drill-press table; *P*, table clamp; *Q*, handle for table jackscrew.

To minimize misalignment, a machine for drilling small holes should be made so that the eccentricity between a chucked drill and the axis of rotation of the spindle is limited to 0.0001 in. There must be no longitudinal looseness or "drop" in the spindle. Small drills with shanks larger in diameter than the cutting blade should be made so that the eccentricity between the shank and blade is less than 0.0001 in.

General-purpose drilling machines, particularly those which have "seen" much service, rarely, if ever, meet these requirements. It is futile to attempt drilling deep, small-diameter holes with such machines.

Sec. 9·9   STARTING A HOLE   119

The Taylor Manufacturing Company (Milwaukee, Wisconsin), Louis Lenin & Son (Los Angeles, California), National Jet Company (Cumberland, Maryland), The Hamilton Tool Company (Hamilton, Ohio), Teletronics Laboratory (Westbury, New York), and others [6] furnish machines intended for drilling small holes.

Figure 9·7 shows a drilling-machine head mounted on a special base suitable for handling work pieces the size of aircraft-engine parts. The *compound-table, adjustable-angle-tilting-table,* and *rotary-table* combination permit precision mounting of the work piece in any position. The micrometer screws make it possible to move the work piece through measured angles and distances. In drilling holes of depth greater than the length of the spindle travel, the table (see O on Fig. 9·7) must be lowered by means of the jackscrew (see Q on Fig. 9·7) to remove the drill from the work.

Figure 9·8 shows mounting arrangements for a small work piece, i.e., a spark plug. A special tilting holder is mounted on a *universal compound vise.* The angle of tilt can be measured with a *universal bevel protractor* held against the table F and the locating plate D on Fig. 9·8.

9·9   STARTING A HOLE

The measured location or point at which a hole is to be drilled should ordinarily be marked by, or *laid out* as, the intersection of two *scribed* lines. The *scriber* is a sharply pointed strip of hardened tool steel mounted in a convenient handle. It should be used somewhat like a draftsman's lead pencil for cutting fine lines on the metal surface. Such lines can be scribed accurately to within 0.001 in.

All distances are measured vertically from the drilling-machine table (see O in Fig. 9·7) using a *vernier height gage.* Thus, after one line has been scribed on the work piece a measured distance from the table, the work piece is rotated 90° and a second line scribed at right angles to the first, likewise a measured distance from the table. The movable jaw of the vernier height gage can be used as a straightedge. In order to relate distances from the table to such dimensions as may be shown on the blueprint of the work piece, it is usually necessary to make preliminary measurements from the table to various machined surfaces on the latter.

When starting to drill a hole, careful procedure is required to insure that the centerline of the hole will pass through the precise intersection of two scribed lines. It is necessary first to ascertain that the work piece is clamped rigidly in position. The *drill locator* (see Fig. 9·9) is then mounted in the chuck of the drilling head, which should be swung into position so that the point of the drill locator is exactly at the center of the intersection of the scribed lines. The sense of touch is helpful for this purpose, since these lines are V-shaped grooves which

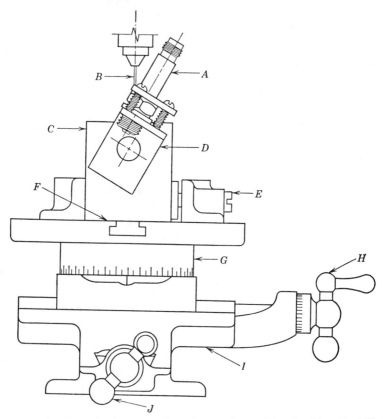

Fig. 9·8. Small work piece on universal mounting. *A*, work piece; *B*, drill; *C*, holding fixture; *D*, locating plate; *E*, vise; *F*, rotary table; *G*, dial; *H*, micrometer screw; *I*, compound table; *J*, micrometer screw.

tend to center the drill locator while under light pressure at the drill spindle. Primary reliance should, however, be placed on direct observation through a strong magnifying glass. When the drill locator appears to be precisely at the intersection, the drilling head should be clamped rigidly in this position. Usually the act of tightening the clamp displaces the head slightly so that further adjustment is required until the drill locator still appears exactly in position after the head has been rigidly secured. An alternative to this procedure is to use an *optical drill chuck*.[7]

The drill locator should then be replaced with a ³⁄₆₄-in. *countersinking tool* having carefully balanced lips. Using the highest available spindle speed and the lightest possible pressure, cut a depression having a flat bottom the diameter of the chisel edge.

The drill locator should then be chucked. At the highest available spindle speed lower the spindle so that the point of the drill locator *spins* a 0.01-in.-diameter conical depression in the metal. Replace the

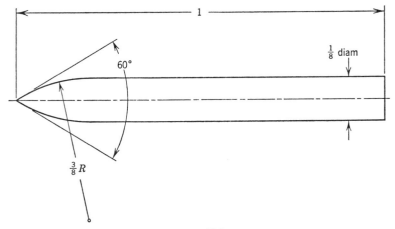

Fig. 9·9. Drill locator.

drill locator with a drill of the required size, if this is not greater than No. 76 (0.020 in.). At the highest available spindle speed and with the lightest possible drill pressure, drill to the required depth. If the required size is greater than No. 76 (0.020 in.), start the hole with a No. 76 drill proceeding to a depth of about 0.062 in. Then replace the first drill with successively larger sizes, drilling deeper each time the drill is changed, until the proper size is reached for the diameter of the desired hole. Three or four stages are usually sufficient.

A drilling machine (see Sec. 9·7) is usually provided with a micrometer-screw stop which can be set to limit the hole to the proper depth. For precise control, however, a special *stop collar* (see Fig. 9·10) is mounted on the shank of the drill itself. Using a *vernier depth gage*, the distance E on Fig. 9·10 from the stop collar to the end of the drill can be set to within 0.001 in.

The hole is drilled to a few thousandths of an inch less than the full depth. The drill is then replaced by a special *square-end reamer* cut from high-speed (HS) steel drill rod to the dimensions indicated on

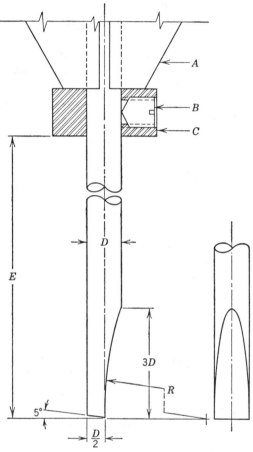

Fig. 9·10. Square-end reamer with stop collar. *A*, chuck; *B*, set screw; *C*, stop collar.

Fig. 9·10 and hardened (see Sec. 9·5). When the stop collar rides gently on the metal at the mouth of the hole, the square-end reamer will have cut the depth of the hole accurately to the distance *C* on Fig. 9·1.

Before the final depth of a hole is measured, all chips must be completely flushed out by means of carbon tetrachloride injected at the

Sec. 9·9 STARTING A HOLE 123

bottom of the hole through a hypodermic needle (see Fig. 9·11). The size required will depend on the diameter and depth of the hole.

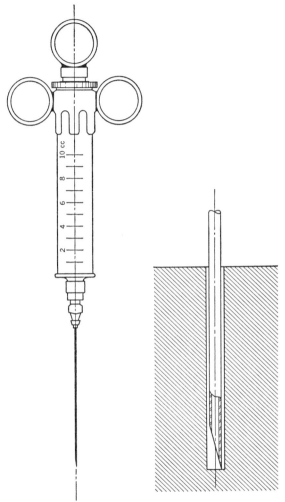

Fig. 9·11. Hypodermic needle.

Between these flushings, a twist drill inserted and moved about lightly by hand will help dislodge the chips.

To check the straightness of the hole, insert a piece of drill rod longer than the depth of the hole and slightly smaller in diameter with rounded ends. Failure of the rod to drop freely to the bottom of the hole should be taken as evidence that the hole is not straight. Accordingly, it would have to be assumed that dimension D of Fig. 9·1 is substantially larger than zero. Such a hole should be rejected.

When the rod rests on the bottom of the hole, the projecting length can be utilized to determine the depth of the hole. Measure the projecting length with the vernier height gage, subtracting the reading to the mouth of the hole from the reading to the projecting end of the rod. This difference, subtracted from the length of the rod as measured with a *micrometer caliper*, gives the depth of the hole.

9·10 DRILLING HARD OR ABRASIVE MATERIALS

The preceding methods apply to drilling in metals such as brass, bronze, copper, aluminum, iron, tool and alloy steels in the annealed condition, wood, fiber, plastics, talc or soapstone, etc. When very hard or abrasive materials are to be drilled, special methods must be employed. Such materials can be classified according to their degrees of hardness.

For drilling materials such as glass, Portland cement, pottery, porcelain, hardened carbon tool steel, hardened high-speed (HS) tool steel, steel armor plate, etc., drills with cutting lips made of *cemented carbide* are used. The cemented carbides of tungsten, titanium, and tantalum in various combinations, usually sintered with cobalt, provide a range in hardness at the expense of toughness. These materials are available commercially under different trade names in a wide variety of grades, shapes, and sizes, including pivot-point microdrills and rotary files similar to dentists' drills. Drills and rods are also available of tungsten-molybdenum-vanadium alloys and of hard alloys of the rare metals. These alloys are tougher but less hard than the cemented carbides.[6] Small drills are made with diamond cutting points embedded in the tips.[8]

The cemented carbides are not only hard, they are also brittle. They cannot be annealed and do not require hardening. Grinding is performed, wet or dry, on aluminum-oxide, silicon-carbide, or diamond-grit wheels. They can be silver-soldered and brazed, suitable fluxes being recommended by the makers of the specific brands.[6]

Pivot-point drills (see Fig. 9·3) and square-end reamers (see Fig. 9·10) can be ground from rods of cemented carbide or of the hard alloys of pure metals. Because of its brittleness, only very short lengths in the cemented carbides are feasible.[1] The hard alloys of rare metals can be used in drills of much greater length-to-diameter ratio. Long, slender drills with cutting edges of cemented carbide must be of the extended-shank variety (see Fig. 9·4). Thus, a short length of cemented carbide is silver-soldered to the end of a length of drill rod and subsequently ground to a pivot point or square-end-reamer point (see Figs. 9·3, 10).

9·11 LAPPING

The *lapping method*, which is costly in time, must be used for drilling in very hard materials such as tungsten, cemented carbide, emery, alumina, silicon carbide, jewels, and such very hard ceramics as those which are used for spark-plug insulators.

The procedure for lapping requires the use of a brass or cold-drawn steel *lapping rod* in the drilling machine instead of a drill. For holes 0.01 to 0.05 in. diameter, this rod is made 0.005 to 0.007 in. smaller than the diameter of the desired hole. If it is necessary to use the lapping method to extend a hole previously started by other means, the rod used is in this case also 0.005 to 0.007 in. smaller in diameter than the hole.

In operation the lapping rod, chucked in the rotating drill spindle, is alternately raised and pressed lightly against the bottom of the hole. A 4-second cycle is satisfactory, allowing 1 second to raise and lower the rod, and 3 seconds for cutting. This motion can be made automatic (see step-drilling attachment, G and H, on Fig. 9·6). It serves to "pump" a fresh suspension of grit to the point of action under the end of the lapping rod.

For small-diameter holes, i.e., less than $3/32$ in. diameter, 180-mesh cutting grit is used. Commercial diamond dust is required for drilling in tungsten, cemented carbide, jewels, emery, silicon carbide, alumina, and such ceramics as those found in spark-plug insulators.[6] For drilling in softer materials such as glass, hardened carbon tool steel, hardened high-speed (HS) steel, and steel armor plate, 180-mesh silicon carbide can be used.

Whether the grit is diamond dust or silicon carbide, it should be suspended in olive oil and applied periodically at the mouth of the hole by means of a toothpick or small piece of wire. Usually it is sufficient to replenish the grit at 15-minute intervals of drilling. The suspension of grit in oil then flows down to the bottom of the hole through the 0.005- to 0.007-in. clearance between the lapping rod and the diameter of the hole. This clearance must at least be sufficient to permit passage of the particles of grit.

9·12 REMOVAL OF BROKEN DRILLS

A drill may be broken off in a hole as a result of an accident, faulty material, or failure to observe suitable precautions. Usually in the breakage the amount of the shank that remains projecting is insufficient to permit extraction of the fragment by direct mechanical means. The work piece must then be scrapped, the position of the hole altered, or means employed for removal of the broken drill. The following are possible methods of extraction: (1), the electric processes; (2), drilling with hard alloys or cemented carbides; and (3), lapping. The choice among these alternatives will depend on the circumstances.

The fragment is removed by *electrochemical methods* if the material being drilled is nonferrous: e.g., aluminum, brass, bronze, copper, or stainless steel. This method involves the gradual dissolving of the broken-drill fragment into the solution. With a small work piece, this is done by submerging the entire piece in the proper solution for the material. For a large body of metal, a glass tube should be sealed over the broken drill and filled with the proper solution. One lead from the power unit should be connected to the work piece, and the other to an anode immersed in the solution. The appropriate solution for a given material will not affect the work piece or alter its dimensions.

In the *spark-drilling method* a vibrating water-flushed electrode, replaces the usual drill. The broken-drill fragment is worn away in the intermittent electric discharge between this electrode and the fragment.

The methods for drilling hard materials can be utilized to remove the drill fragment. Of these, lapping is certain and universally applicable, but costly in time. In drilling with hard-alloy or carbide-tipped drills, the fragment, being harder than the parent material, tends to deflect the drill. This effect can be lessened by using extremely light pressure. If such runout is observed to begin, lapping can still be resorted to.

REFERENCES

1. I. H. Such, "Microscopic Precision Tools," *Steel*, 113, no. 11, pp. 98–100, 102 (Sept. 13, 1943).
2. L. G. French, C. L. Goodrich, J. M. B. Scheele, F. B. Kleinhans, O. Eckelt, and E. W. Norton, "Deep Hole Drilling," *Machinery's Reference Series*, no. 25, Industrial Press, New York, 1908.
3. W. W. Gilbert and A. M. Lennie, "Drilling Deep Holes in Magnesium Alloys," *Mech. Eng.*, 64, no. 12, pp. 877–887 (December, 1942).
4. F. O. Hoagland, "Grind Deep-Hole Drills Properly," *Am. Machinist*, 87, no. 4, pp. 91–93 (Feb. 18, 1943).

REFERENCES

5. "Drilling Accurate Holes of Small Diameter," *Ind. Power & Production*, **19**, pp. 111–113 (May, 1943).
6. American Society of Mechanical Engineers, *A.S.M.E. Mechanical Catalog and Directory*, **42**, pp. 450, 457–459, New York (1953).
7. "Optical Drill Chuck," *Mech. Eng.*, **69**, no. 9, pp. 775–776 (September, 1947).
8. "Diamond Drills," *Rev. Sci. Instr.*, **22**, No. 8, p. 658 (August, 1951).

10

SPECIAL MATERIALS: PROTECTIVE COATINGS, HEAT- AND CORROSION-RESISTANT METALS, PLASTICS, REFRACTORIES, AND CEMENTS

10·1 INTRODUCTION

Frequently, the circumstances in which temperature is measured require that portions of the instrumentation function under difficult conditions. Some measurement methods necessitate that a section of the instrumentation attain a temperature approximately equal to that being measured. Portions of the instrumentation may be subjected to severe chemical conditions, such as exposure to corrosive gases or liquids, or high humidity prevailing in the zone in which temperature is to be measured. Mechanical vibrations may also be a factor. The materials of which the structure is made, as well as its form, must both be adapted to the specified conditions for a suitable operating lifetime.

Functions required to be performed by individual elements of the structure, and for which the materials must be in conformity, may include maintenance of electrical resistance, conductivity, or insulation; thermoelectric power; reflection, transmission, or absorption of radiation; thermal conduction or insulation; and mechanical strength or rigidity. Materials possessing otherwise suitable properties can be utilized only if some method is available for their fabrication into a suitable structure.

The scope of available materials is almost unlimited. New varieties are announced in each issue of the technical journals (see Refs. 1 to 7, Ch. 6). The present chapter deals with a very few of these, which have been found to be useful in the personal experiences of the authors. Those optical properties which are pertinent for measuring temperature by radiation methods are deferred for discussion to Vol. II.

10·2 PROTECTIVE COATINGS

The electric circuit of a thermoelectric thermometer or electric-resistance thermometer must be properly insulated against electrical leakages.

In situations where these methods of temperature measurement are applicable, the most reliable insulation is likely to be attained by thoroughly drying the wires and then dipping them into melted *paraffin*. Insulation of this sort can be given additional mechanical strength by silk or cotton windings on the wires. At best, however, paraffin withstands only a small amount of mechanical abuse. Moreover, it cannot be operated at much above room temperature. Nevertheless, in circumstances where the highest degree of insulation is required in a humid environment, under water, or in the presence of certain chemicals, paraffin should seriously be considered as an insulator.[1]

Natural *ceresin* wax has high electrical resistivity and extraordinarily low surface conductivity, even under conditions of high relative humidity. It does not soften at quite as low temperatures as paraffin, and is, therefore, rigid at room temperature in the hottest weather.[1,2]

Beeswax in the pure state is tougher and more adherent than paraffin or ceresin, but of lower electrical resistivity. It can be applied dissolved in benzene. Even after being vigorously rubbed down with a cloth, a sufficient film remains to prevent rust or other corrosion at room temperature in a humid atmosphere for steel and similar materials. The toughness of beeswax can be increased by adding dissolved rosin and gum rubber. On the other hand, addition of dissolved vaseline tends to make it softer.[1]

By varying the proportions of vaseline, beeswax, rosin, and gum rubber, greases or waxes of a continuous range of stiffness can be obtained, with vaseline at one extreme and beeswax, rubber, or rosin at the other. A solution of these materials can be obtained by warming them together at a temperature somewhat above the melting points of vaseline and beeswax.[2]

The highest-quality spar varnishes withstand boiling water for a brief time and high humidity or cool water for long periods, but when wet do not provide as good insulation as paraffin, ceresin, or beeswax. They can be applied by brushing, spraying, or dipping and are air-drying. They can be removed with ordinary paint remover and withstand much mechanical abuse.[3,4]

A wide assortment of plastic-base coatings is available, from which those with the required properties can be selected. These include both air-drying and thermosetting types. For example, lacquers based on the *alkyd resins* (Glyptal), available from the General Electric Company (Schenectady, New York), are furnished in forms which, if cured by baking after preliminary air-drying, have low vapor pressure and high electrical resistance and withstand hot water, acids, alkalies, and gasoline. These are rated for service at up to about 220°F, but often have utility at considerably higher temperatures (see Sec. 11·4).[2-6]

For temperatures above those that essentially organic materials can withstand, the silicones are available. There are various varnishes and paints requiring, in general, to be cured by baking. The usual highest-recommended operating temperature is 500°F or below; however, certain of the paints have been found to withstand temperatures of 500 to 1000°F.[4-6]

To achieve its best properties as an insulating coating, a varnish or lacquer should usually be baked in accordance with the manufacturer's directions. The particular baking process to be used may, however, also depend on the proposed application. Thus, in general, it is desirable to bake any varnish at a temperature somewhat above that at which it is to function in service. If a temperature lower than the operating temperature is used for baking, the varnish may tend to soften during the initial period in operation. If the varnish alone is depended upon to maintain spacing between electric circuits, it may be squeezed out during this first period in use or during baking, with a resultant failure in insulation, i.e., a short circuit.

Except where joints must be made or other circumstances forbid, it is desirable to use ready-insulated wires such as are obtainable from various commercial sources. Materials used for insulation on wires of this type include silk, cotton, waxed cotton, rubber, asbestos, enamel, mixtures of powdered ceramic and plastic, and fiber glass impregnated with varnish. The manufacturer's specifications should be consulted to ascertain the properties of a given insulator under various operating conditions. The insulation that comes with the wire may often advantageously be supplemented with applications of wax or varnish.[2-6]

In choosing an insulator, consideration must be given to the probable operating temperature. Organic materials such as silk, cotton, rubber, and the various plastics generally tend to fail in the temperature range 200 to 350°F. Fiber-glass windings can be used up to 650 or 750°F, depending on the required insulation resistance. Silicone varnishes, enamels, and paints withstand temperatures up to 500 and 1000°F. Added strength and reliability can be obtained by using fiber glass impregnated with silicone. In general, all electrical insulators tend to become more and more conducting of electricity at higher temperatures. Glass becomes a semiconductor of electricity at temperatures above 800°F.[2-8]

A number of commercially available varnishes, sprays, and impregnations are intended to render surfaces *water repellent,* i.e., such as not to be *wetted* by water. Complete short-circuiting through a continuous water film is thereby prevented, when these are applied to electrical insulations.[4-6, 9]

10·3 HEAT- AND CORROSION-RESISTANT METALS

Although electrical circuits must ordinarily be made of metal, a choice of plastics, ceramics, or metals can be utilized for mechanical structural elements. Metals, however, are much more reliable than plastics or ceramics, from the standpoint of both strength and resistance to mechanical and thermal shock.

Circuits supported on metal structures can be provided with electrical insulation by (1), putting an insulating coating on the wires; (2), encasing the wires in insulating tubing, i.e., "spaghetti"; or (3), separating the circuit metals from the structural metal by interposing layers of insulating material in the form of sheets, gaskets, washers, bushings, or blocks of other shapes.

Metal used for structure alone can usually be protected from the effects of a high-temperature region, if water-cooled by means of an internal-circulating system. Structures cooled in this way can be exposed to the hottest gases and to the most intense radiations occurring in practice. This arrangement permits utilization of easily fabricated metals, such as brass joined by either soft or hard solders.

If it is considered desirable to dispense with water-cooling, the structure must be made of materials which will withstand the corresponding thermal conditions. In an atmosphere that furnishes protection against chemical attack, a given metal can be used up to the temperature at which its stress resistance becomes inadequate to support the imposed mechanical loading, or at which its rate of dissipation by evaporation becomes excessive.

The environment is usually, however, not subject to control to afford protection against chemical attack. An ambient atmosphere commonly consists of either air or the products of combustion of coal or other hydrocarbon fuels. A liquid environment may consist of neutral substances, water, or chemicals which are corrosive even at room temperature or below. When selecting a metal for structural purposes, resistance to specified chemical conditions must therefore often be considered a determining factor.

Structural metals possessing varying degrees of inertness in a given chemical environment are available at different degrees of costliness. Among such metals, platinum does not oxidize at any temperature up to its melting point, 3250°F, and is attacked by very few chemicals in liquid solution. It should not, however, be heated in the presence of oxidizable materials. Metallic salts, particularly lead salts; carbon, sulfur, phosphorus, and their oxidizable compounds; and silicon and silicon compounds, such as porcelain, mica, and soapstone, should not be heated in contact with platinum (see Secs. 5·5, 8). The result of such treatment would be embrittlement of the platinum and alteration in others of its properties, such as thermoelectric power and electrical resistance. There seems to be no difficulty in operating platinum heater filaments on fused-alumina supports in air at 3000°F. Platinum dissolves in aqua regia, a mixture of hydrochloric and nitric acids.[2, 4, 9–12]

A number of other precious and semiprecious metals and metallic alloys which are highly inert in common liquids have been developed in the dental and food industries. These include gold, both pure and alloyed, and tantalum. Tantalum exhibits the additional desirable property of a very high melting point, i.e., 5425°F.[2, 4, 9–12]

As a metal to be used in water, ordinary brass stands up almost indefinitely. To provide greater mechanical strength and elasticity, the bronzes can be utilized instead of brass. For additional protection against water, nickel or chromium plating can be added to brass, bronze, and other metals. Pure nickel can also be used in water. The natural proportions, as mined, i.e., approximately 67 per cent nickel, 30 per cent copper with small amounts of iron, manganese, silicon and carbon, provide an alloy (Monel) widely used for its corrosion resistance combined with strength and toughness. This alloy is cheaper than pure nickel.[2, 4, 11, 13]

Among the commercial iron-nickel alloys the *stainless steels* (commonly 18 per cent nickel, 8 per cent chromium with the remainder consisting mostly of iron) are recommended for their strength and cheapness. They do rust, however, with a speed depending on conditions. An alloy consisting of approximately 36 per cent nickel and 64 per cent iron (Invar) is useful where thermal expansion must be minimized. It is strong and tough, but rusts slowly, at a rate depending on the circumstances. Another alloy, consisting approximately of 78 per cent nickel, 14 per cent chromium, and 6 per cent iron with small amounts of copper, manganese, silicon, and carbon (Inconel) is often useful because of its extraordinarily low thermal conductivity, i.e., 8.7 Btu/hr ft°F which is about the same as that of silicon carbide. It is gastight, readily machined, resistant to reagents and high-temperature corrosion, and is strong and tough. Various alloys, which consist, approximately, of 60 per cent nickel, 24 per cent iron, 16 per cent chromium and 0.01 per cent carbon (Nichrome, etc.), are used when operating conditions involve exposure to air or products of combustion at elevated temperatures. Pure nickel is strong, tough, stainless, and useful exposed to air at elevated temperatures.[2,4,11–13]

These nickel-iron alloys can all be soldered, silver-soldered, brazed, and welded, although those containing chromium require special fluxes. Feasible operating temperatures range from 1800 to 2200°F, depending upon the composition of the alloy, the exact composition of the environmental gases, and the required lifetime in service. Stress resistance is low at temperatures near to the melting points.[2,4,11–13]

Molybdenum, which melts at 4760°F, and tungsten, which melts at 6170°F, are stiff and strong at relatively high temperatures, but require protection from oxygen.

At temperatures above 3000°F, molybdenum rapidly discharges its vapor, which, condensing elsewhere, tends to "short out" electrical connections. If protected by an inert atmosphere, such as helium or argon or by a suitable ceramic-enamel coating to prevent corrosion, molybdenum can be used structurally at temperatures that would otherwise necessitate the use of ceramics. As furnished commercially, it is a tough, ductile metal which can be cut, formed, machined, or drawn in dies by ordinary methods. This ductility depends on purity, and on the mechanical and heat treatments applied in manufacture. It is difficult to produce ductile welds in molybdenum because contamination and recrystallization cause embrittlement. It can be spot-welded if the proper circuit arrangements are employed. It can also be welded by using atomic-hydrogen or helium-shielded arcs. Suitable resistance-welding machines and ready-made atomic-hydrogen torches are available commercially.[2,4,11,12,14,15]

Tungsten stock, in the form in which it is usually supplied, is too brittle at room temperature to be bent or formed, although ductile wire is produced in small diameters. It is also too hard to be machined except by the use of cemented-carbide tools or by grinding. As a result, presumably, of both recrystallization and temperature stresses, tungsten is very difficult to weld, and methods not available in the ordinary shop must be used. Tungsten can be used in self-supporting heater filaments at temperatures up to 5500°F, without excessive discharge of vapor.[2,4,11,12]

In general, and particularly in heater filaments, refractory metals will withstand higher temperatures and more severely corrosive conditions in uniform and heavy cross sections. "Thin spots" tend to become "hot spots." This is because of increased thermal resistance along the paths leading to cooler zones in the material and, also, in the case of heater filaments, to increased local Joule heating. In heavy sections, not only is the reserve material more copious, but relative uniformity in thickness is not so readily upset by initial surface irregularities and wearing away of the metal in service.

10·4 PLASTICS

For purposes of structural fabrication, plastics have certain desirable properties not possessed by metals. They display excellent characteristics as electrical insulators and also have some value as thermal insulators. They resist most chemicals better than metals, and are affected but slowly by water and various hydrocarbon liquids. In common with metals, plastics have the property of being unaffected by thermal shock. On the other hand, their stress resistance is usually low, and operating temperatures must be closely limited to within the permissible ranges.

Plastics and similar natural materials, such as amber, are therefore used for electrical insulating members wherever temperatures permit. Under humid conditions, where such insulation is depended upon, precision electrical measurements are difficult because surface condensation of moisture causes excessive conduction of electricity. To offset the effect of humidity, special measures may be necessary, such as replacement by, or coating with, natural ceresin wax or other water-repellent coating; immersion under oil; or provision of drying arrangements by some form of air-conditioning apparatus to dry the air in the room.[1,2,6,9]

The temperature at which heat distortion occurs ranges from 120 to 190°F for most of the plastics; whereas the upper temperature limits are from 250 to 350°F. Polytetrafluoroethylene (Teflon), available from E. I. duPont de Nemours & Co., Inc. (Wilmington, Delaware), however, can be used at temperatures up to around 550°F; it is strong, impervious, tough, readily machined, and resistant to nearly all chemicals and solvents. Several combinations of mica and fiber glass with the silicones are available. These can usually be used at temperatures up to around 500°F, although 1000°F silicones are available.[2,4,5,8]

Gaskets to seal joints against fluid leakage can be made of natural rubber, synthetic rubber, leather, textiles impregnated with rubber, or plastics of various degrees of softness. A material that can be utilized for this purpose at temperatures up to 500°F is a combination of mica and fiber glass impregnated with silicone. Asbestos paper or fiber-glass fabric coated with silicone rubber can be used over the range from −100 to 500°F.[2,4,5,9]

Insulating tubing called "spaghetti" can be utilized to provide electrical insulation for otherwise bare wires. The "spaghetti" is made of porcelain, or of cotton or glass braid impregnated with impervious, flexible dielectric material such as the various plastics and silicones.[4-8]

10·5 NONMETALLIC REFRACTORIES

Nonmetallic refractories, particularly *ceramics*, are often useful for purposes of electrical and thermal insulation under circumstances where operating temperatures are above those permissible for plastics. They are also useful for structural members where temperatures and possibilities for protection from corrosive materials are such as to prevent the utilization of metals.

Nonmetallic refractories tend to be weak, brittle, porous, easily destroyed by thermal shock, impossible to weld, and impossible to machine; however, there are notable exceptions to each of these tendencies. They can be joined effectively by cements. At room temperatures they are quite inert; however, at the elevated temperatures, where they are most likely to be useful, they are subject to a variety of chemical hazards. With the exception of carbon, nonmetallic refractory materials are usually electrical insulators, but at elevated temperatures many become semiconductors. Since a large number of materials are commercially available, the properties of which vary widely, the manufacturer's specifications should always be consulted.[2,4,7,8,16,17]

Among the useful ceramics are the various forms of artificial corundum. These consist of granular alumina bonded with suitable clays. Depending on the amount and composition of the bonding clay used, these materials withstand temperatures up to 3000 or 3400°F, provide fair electrical insulation at such temperatures, are moderately strong, can be molded in a variety of shapes, are impervious to gases, and do not rapidly contaminate platinum or platinum-rhodium thermocouple wires. Cements, which cure to similar substances, are available. Joints or entire parts can readily be made from certain of these cements. Firing at a temperature of 2000°F or higher is usually required.[2,4,16,17]

The high-alumina ceramics used for spark-plug cores are extraordinarily strong, hard, and resistant to mechanical and thermal shock.

Granular silicon carbide bonded with suitable clays is sold under various trade names for use at temperatures up to 3000 or 3200°F. A glazed form is said to be impervious to gases. When the material is nearly pure silicon carbide, the thermal conductivity is much higher than that of most nonmetals, i.e., about 9.1 Btu/hr ft°F. Electric-resistance heater elements for use at temperatures up to around 3000°F are made of a semiconducting ceramic material consisting essentially of silicon carbide.[2,4,16]

Porcelains consist of fired clays. Depending on the composition and grain sizes in the clay, and on the manner of molding and firing, a wide variety results. Most porcelains are intended for use at temperatures not exceeding 2100 to 2500°F; however, porcelains are available that withstand 3000°F. Certain porcelains permit fabrication in exceptionally small sizes, complex shapes, and thin walls. There are varieties impervious to gas.[4,17]

Synthetic mullite is furnished in forms highly resistant to thermal shock. Talc, i.e., soapstone, before firing, is soft and readily machined. The iron-free or "radio-grade" talc should be used. Subsequent firing at 1300 to 1800°F for several hours causes hardening, with removal of the water of crystallization and consequent shrinkage with possible warping. Before firing, steatite can be machined with great precision, using cemented-carbide cutting tools. Subsequent firing causes very small dimensional changes and results in a ceramic which can be used at temperatures up to 2250°F.[4,9]

Ordinary glass is a good electrical insulator at temperatures up to around 650°F and is inert chemically. It offers the advantage of easy fabrication into a wide variety of shapes by "glass-blowing" technique. "Sealed" joints can be made between glass and metals. Although, in general, stronger and tougher than most other nonmetallic refractories, glass tends to become soft somewhat before reaching a dull red heat and shatters if subjected to severe thermal shock. The various "heat-resistant" glasses, often consisting essentially of *fused silica,* are much more resistant to thermal shock than ordinary glass and do not soften at quite such low temperatures.[2,4,7,8,9,18]

Fused silica is unharmed by the most extreme forms of thermal shock. Various brands are recommended for continuous service at temperatures up to 1800°F, or, for short periods under suitable conditions, at up to 2700°F. Fused-silica ware is available in a large variety of shapes and sizes. It is readily cut by diamond-charged wheels.[2,4,7,8,9,18]

Mica is widely used for electrical insulation at temperatures above the permissible limit for plastics and the natural resins. For example, the coils in precision electric-resistance thermometers, specified under the International Temperature Scale for the range -297.35 to $1166.9°F$, are wound on mica. Samples from the various natural sources differ widely in properties. When raised to elevated temperatures, mica loses its water of crystallization, becomes opaque, and suffers dimensional changes up to and exceeding ± 25 per cent. Few samples fail to retain their utility up to 1100°F, whereas 1800 to 2000°F is the upper limit. The synthetic variety is said to withstand higher temperatures.[2, 4, 8, 9, 19, 20]

In the bonded ceramics of alumina, silicon carbide, etc., the operating temperatures are limited by the softening points of the bonding clays to 3000 or 3400°F. For higher-temperature work no bonding agent is used, the pure materials being fused together directly. The limiting useful temperature is then close to the melting point of the pure material used. Thus, alumina 99.7 per cent pure is available in a broad assortment of ware intended for service temperature up to 3540°F. Fused magnesia, 95 to 96 per cent pure, is said to be stable in oxidizing or neutral atmospheres at temperatures up to 4530°F. Thoria, 99.9 per cent pure, melts at 5430°F, is stable in oxidizing and neutral atmospheres, and is not affected by carbonaceous atmospheres up to 4000°F. Zirconia, 96 to 97 per cent pure, is said to be suitable for use in oxidizing or moderately reducing atmospheres at temperatures up to 4350°F. Zirconia, fused with additives intended to increase dimensional stability, is said to resist thermal shock, have exceptionally low thermal conductivity, and remain serviceable at temperatures approaching 4600°F. These materials can be cut with diamond-charged wheels.[4, 16, 21–23]

Pure zirconium silicate fired at its melting point is said to be serviceable and gastight up to 2900°F. Its low thermal coefficient of expansion, i.e., about one-third that of pure silica, renders it extraordinarily resistant to thermal shock.

Fibers of fused alumina and silicon carbide do not soften below 3000°F and are considered serviceable up to 2300°F.

Carbon is a good electrical conductor and is available in a large number of accurately formed shapes and sizes. It oxidizes when heated in air, but does not melt at any temperature thus far artificially achieved.[2, 4, 6]

10·6 CEMENTS

Since cemented joints are usually relatively weak, a joint should be made with cement only as a last resort. If properly designed, joints that are welded, brazed, soldered, screwed, riveted, or otherwise mechanically locked, can reasonably be expected to withstand temperature changes, vibration, mechanical and thermal shocks. The objections to cemented joints are less valid for weak materials, such as certain ceramics and plastics, than for metals, because with nonmetals the strengths of the cemented joints and of the parent materials are more nearly equal. Moreover, the greater similarity in thermal conductivity tends to increase the resistance of the joint to thermal shock.

In an intricate structure, however, satisfactory joints locked by metal fastenings are frequently impossible to achieve due to space limitations or other circumstances. In a number of such cases skillful use of cement provides the solution to the fabrication problem.

A large variety of cements is available, each with its own field of usefulness. These include such diverse substances as common glue, rubber cement, the various waterproof resins, synthetic plastic products, dental and ceramic cements.[1, 2, 4-6, 9, 17, 24, 25]

On the basis of method of application, cements can be classified as (1), *melting* or *thermoplastic;* (2), *air-drying;* (3), *thermosetting*, i.e., requiring to be cured by application of heat and, perhaps, also pressure; and (4), *chemically setting.*

The melting cements are usually waxes with various admixtures to provide extra stiffness or toughness. They generally come in the form of sticks roughly the size of a lead pencil. The end of a stick is often merely melted in a small flame and applied to the object to be cemented. Where possible, however, it is preferable to heat the objects to be cemented and apply the stick of cement to the heated surfaces. Cements of this type are ordinarily limited to service at room temperature or below. The most common variety of melting cement is the traditional red sealing wax. The half-and-half mixture of rosin and beeswax is notably adherent and tough. Shellac tempered with wood tar (De Khotinsky and Sealstix), available from Central Scientific Company (Chicago, Illinois), adheres well to most rigid bodies and resists a wide variety of reagents and solvents but dissolves in alcohol. It is very strong and tough.[1, 2, 4, 6, 9, 24]

The air-drying category includes most of the household cements. The drying process requires that the liquid, in which the cementing material is dissolved, must diffuse to the surface of the layer of cement and be removed, either by evaporation or by soaking into the materials being joined. Drying involves considerable time, even under the most favorable circumstances. Cements of this type can be used only in thin layers and for joining porous materials; they are usually limited to temperatures below 200 to 300°F.[2, 4-6, 9, 24]

Among the group of cements that require the application of heat for *curing* are included certain ceramic cements for use at elevated temperatures. Usually the temperature, when setting, must be sufficiently high to melt or soften certain of the constituent materials, which then serve as binders for the others. The rule is to fire, if feasible, at as high a temperature as any to be encountered in operation.[4, 17, 18]

Several varieties of plastics, as, for example, the kind used in making plywood, can be used as cements. These substances are usually *thermosetting*, i.e., they must be cured or polymerized by being molded together with the parts being joined at the required temperature, often also, under heavy pressure. Thermosetting silicone cements are usable at temperatures up to 475°F. The cured plastic cements have very great strength but are usually practicable only for mass production.[2,4,5,9,24]

Unlike the air-drying cements, chemically setting cements can be used in thick masses and in locations confined by impervious materials. By varying the proportions of the constituents, the length of time required for setting can usually be adjusted and sometimes can be reduced to a few minutes. These cements are commonly mixed immediately before use. The various dental cements belong to this category; most of these substances are limited to service at 100-per cent humidity.[4,17,25] Among the exceptions to this limitation is Technical "B" Copper Cement, available from the W. V-B Ames Company (Fremont, Ohio). This cement retains most of its strength after heating (dry) to 1400 or 1600°F, adheres well to metals, resists mechanical and thermal shock, and has unusual strength and toughness. The mixture of liquid and powder makes a highly homogeneous, creamy paste, which can be worked to the bottom of a hole several inches in depth and as small as $1/64$ in. in diameter. Any trapped air can be eliminated by moving up and down, while rotating, a wire of about half the diameter of the hole. If mixed to a suitable consistency, this cement sets in about 15 min.

A number of chemically setting porcelain cements are available. These cements are intended for service at temperatures up to 2000°F. Their consistency, however, is usually not so fine and creamy before setting as that of the Technical "B" Copper Cement, thus limiting their application to coarser work.[4,17,25]

REFERENCES

1. H. Bennett, *Commercial Waxes*, 583 pp., Chemical Publishing Co., Brooklyn, 1944.
2. J. Strong, H. V. Neher, A. E. Whitford, C. H. Cartwright, and R. Hayward, *Procedures in Experimental Physics*, pp. 28, 126, 129, 131, 152, 194, 258, 348, 387–390, 520, 521, 542–549, 554–563, Prentice-Hall, New York, 1941.
3. Office of Technical Services, U. S. Department of Commerce, *Bibliography of Reports on Protective Coatings for Metals*, 49 pp., Washington (October, 1947).
4. American Society of Mechanical Engineers, *A.S.M.E. Mechanical Catalog and Directory*, 42, pp. 378, 400, 402, 408, 416–418, 420, 422, 425, 477, 497, 518, 534, 538, 540, 556–558, 576, 602, 655, 660, New York (1953).

5. "Plastics: A List of Books and Periodicals," *Letter Circular* LC 975, GMK: IMD 7·7, 8 pp., National Bureau of Standards, Washington (Jan. 13, 1950).
6. *Radio's Master*, 16th edition; pp. J-13; P-66, 66A; R-9, 21, 33; S-12A, 23, 29, 40, 41, 46; U-69, 85–87, 95, 97, 101, 125; United Catalog Publishers, New York, 1951.
7. H. J. Hoge, "Electrical Conduction in the Glass Insulation of Resistance Thermometers," RP 1466, *J. Research Natl. Bur. Standards*, 28, no. 4, pp. 489–498 (April, 1942).
8. James G. Biddle Co., "Temperature-Resistance Characteristics of Electrical Insulation," *Technical Publication* 21T4, 16 pp., Philadelphia (1946).
9. C. H. Bachman, *Techniques in Experimental Electronics*, pp. 51–60, 68–88, 127, 138–140, 214–224, John Wiley & Sons, New York, 1948.
10. E. Wichers, W. G. Schlecht, and C. L. Gordon, "Attack of Refractory Platiniferous Materials by Acid Mixtures at Elevated Temperatures," RP 1614, *J. Research Natl. Bur. Standards*, 33, no. 5, pp. 363–381 (November, 1944).
11. H. H. Uhlig, *Corrosion Handbook*, 1118 pp., John Wiley & Sons, New York, 1948.
12. F. H. Clark, *Metals at High Temperatures*, 372 pp., Reinhold Publishing Co., New York, 1950.
13. "Nickel and Its Alloys," *Natl. Bur. Standards Circ.* 485, 72 pp., Government Printing Office, Washington (March, 1950).
14. W. N. Harrison, D. G. Moore, and J. C. Richmond, "Review of an Investigation of Ceramic Coatings for Metallic Turbine Parts and Other High-Temperature Applications," *Technical Note* 1186, 17 pp., U. S. National Advisory Committee for Aeronautics, Washington (March, 1947).
15. R. A. Long, K. C. Dike, and H. R. Bear, "Some Properties of High-Purity Sintered Wrought Molybdenum Metal at Temperatures Up to 2400°F," *Technical Note* 2319, 75 pp., National Advisory Committee for Aeronautics, Washington (March, 1951).
16. J. J. Gangler, C. F. Robards, and J. E. McNutt, "Physical Properties at Elevated Temperature of Seven Hot-Pressed Ceramics," *Technical Note* 1911, 33 pp., U. S. National Advisory Committee for Aeronautics, Washington (July, 1949).
17. C. F. Sauereisen, "Choosing the Right Technical Cement," *Elec. Mfg.*, 47, no. 4, pp. 136, 138 (April, 1951).
18. W. E. Barr, *Scientific and Industrial Glass Blowing and Laboratory Techniques*, 388 pp., Instruments Publishing Co., Pittsburgh, 1949.
19. A. B. Lewis, E. L. Hall, and F. R. Caldwell, "Some Electrical Properties of Foreign and Domestic Micas and the Effect of Elevated Temperatures on Micas," RP 347, *J. Research Natl. Bur. Standards*, 7, no. 2, pp. 403–418 (August, 1931).
20. P. Hidnert and G. Dickson, "Some Physical Properties of Mica," RP 1675, *J. Research Natl. Bur. Standards*, 35, no. 4, pp. 309–353 (October, 1945).
21. W. H. Swanger and F. R. Caldwell, "Special Refractories for Use at High Temperature," RP 327, *J. Research Natl. Bur. Standards*, 6, no. 6, pp. 1131–1143 (June, 1931).
22. C. E. Curtis, "Development of Zirconia Resistant to Thermal Shock," *J. Am. Cerm. Soc.*, 30, no. 6, pp. 180–196 (June, 1947).
23. "Fused Stabilized Zirconia," *Mech. Eng.*, 73, no. 6, pp. 507–508 (June, 1951).

24. R. C. Rinker and G. M. Kline, "Survey of Adhesives and Adhesion," *Technical Note* 989, 93 pp., U. S. National Advisory Committee for Aeronautics, Washington (August, 1945).
25. I. C. Schoonover and W. Souder, "Research on Dental Materials at the National Bureau of Standards," *Natl. Bur. Standards Circ.* 497, 14 pp., Government Printing Office, Washington (August, 1950).

11

CEMENTED INSTALLATION DESIGNS

11·1 INTRODUCTION

For the sort of installations described in this chapter and in Ch. 12, thermal contact with the body occurs not only at the sensitive element or junction, but along the length of the leads or thermocouple wires inserted into the material. As a result of this thermal contact, heat flow in the leads is dissipated along the wires for the length of their insertion (see Secs. 7·2, 12). The length of immersion effective for heat transfer, i.e., L in Eq. 7·9, is considered to be the length of the bared wire immersed in the material. The designs described indicate the length of insertion necessary to immerse properly the bared portion; greater lengths of immersion have no significant effect (see Fig. 11·3).[1]

Under these conditions of thermal contact, the temperature of the wires rapidly approaches that of the opposing material of the body at increasing distances from the point of entry into the body. If the body temperature is uniform along the direction of the leads, the discrepancy in temperature between the leads and the opposing body material becomes negligible. If, however, there is a variation of the body temperature, i.e., a *temperature gradient* (see Sec. 12·1) along the immersed part of the leads near to the junction, there will be a final discrepancy in temperature proportional to the magnitude of this gradient. This discrepancy is given by the second term in Eq. 7·9.[1]

The instructions detailed here are phrased in terms of those specific materials with which the authors have had personal experience. Other equivalent materials may be found equally satisfactory.[1]

11·2 LEADS EMERGING TOGETHER

The essential features of a design, which has had a wide range of application in the internal-combustion-engine industry, are indicated on Fig. 11·1. This design, requiring approach from only one direction, can be installed in any solid material. The depth to which the junction

Sec. 11·2   LEADS EMERGING TOGETHER

is inserted into the material may range from ¼ in. to any greater depth, for which the drilling of the required hole in the specified material is feasible by means of available drilling methods.[1]

This installation is sufficiently rugged in the completed form to serve average industrial test-stand requirements. It can be produced from commercially available materials. Its precision in the measurement of temperature will depend on the quality of the thermocouple wire used, on the span between the "hot-" and "cold-"junction temperatures, on the magnitudes of the parasitic emfs, and on the external circuit arrangements. The accuracy with which the location of the temperature measured can be known is generally to within less than 1/64 in. An installation of this design can be used at temperatures up to around 1400°F and at lower temperatures down to about −425°F. As described, it is not intended for highly humid or submarine conditions; suitable coating or sheathing (see Secs. 10·2 and 11·9) of the leads can, however, adapt it to such service.[1]

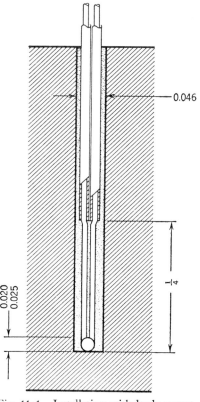

Fig. 11·1. Installation with leads emerging together.

The installation is performed as follows. A No. 56 (0.046 in.) hole is drilled to the required depth in the material, at the correct location and angle. Since the actual junction occurs at the point at which the two wires enter the bead, the hole must extend beyond the point at which it is desired to measure temperature, and by an amount equal to the length of the bead (see Fig. 5·5). The precision of the installation is limited, in part, by the accuracy of the drilling operation. If the work is properly done, the location of the bottom of the hole should be as specified to within a few thousandths of an inch.

Number 30 B & S gage (0.010 in.) glass-insulated, duplex, iron-against-constantan thermocouple wire is used. The outer insulating braid is removed from the length of the wire which will be inside the hole. The inner winding is removed for about ⅜ in. back from one end of each wire. After twisting the ends together, all but about 1½ turns are clipped off and the remainder is melted into a bead (see Sec. 5·8). The length of the bead (see Fig. 5·5) is then trimmed to 0.020 in. by hand-grinding on fine emery cloth. About ¼ in. of bare wire should remain above the bead; all the twisted portion of the wire should have been absorbed into the bead. The distance from the point at which the wires enter the bead to the end of the bead is then measured.

After the bead is made, the bared portion of the wire must be coated to provide electrical insulation. Technical "B" Copper Cement (see Sec. 10·6) is used for a first coat. A second coating of very thin clear Glyptal varnish (see Sec. 10·2) is applied after the cement has set. Care should be taken to have the wires close together but not in contact, except at the junction. The copper-cement coating provides a thermally conducting electrical insulation which will withstand high temperatures. The varnish provides a tough exterior to protect the copper cement during installation of the couples.[1]

The hole is cleaned with carbon tetrachloride (see Sec. 9·9). The copper cement is mixed to a cream and the hole is filled with this cream. By using a wire about half the diameter of the hole as a "piston," it is possible to "work" (or "pump") the cement to the bottom of the hole and to be sure of having it there. When the hole will absorb no more cement, the prepared junction is painted with the copper-cement cream and immediately inserted into the bottom. It is lifted up and down a few times and then held touching the bottom of the hole. In about 15 min the cement will have set. The progress of the setting process is indicated by the gradual thickening of the unused portion of cement remaining in the mixing dish.

When the cement has set, the installation is complete, unless it is necessary to anchor and support the leads at their point of emergence from the body.

11·3 ANCHORING THE LEADS

Figure 11·2 indicates a method of anchoring the leads that has satisfactorily withstood the severe vibrations prevailing when the junction is installed within a cooling fin on a single-cylinder aviation engine. For installation in such an engine, the binding wire can conveniently be secured by passing it through a crosshole drilled in the fin. In other installations, the wire can be secured by a machine screw tapped into the body. A suitable material for the binding is No. 24 B & S gage (0.020 in.) bare wire, preferably of a tough material such as a nickel alloy. The entire joint is finally given a heavy coat of thick G.E. 1201 Glyptal Red Enamel.

Where the leads are *sheathed*, the end of the sheathing must be included in the anchorage.

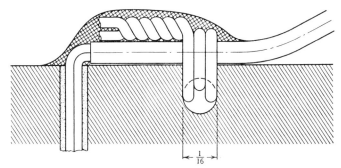

Fig. 11·2. Method of anchoring leads.

11·4 TESTS

The installation described in Sec. 11·2 was made the subject of an extensive series of tests designed to determine its precision and reliability in service, as well as the factors on which these qualities depend. Whereas the installations described in Secs. 11·2; 11·5 to 11·8; and 12·2 to 12·4 all depend on a common principle (as described mathematically in Sec. 7·12), and differ only in scale and materials, it was intended that these tests evaluate all such installations. Specifically, these tests affirm the uniform accuracy with which the precision of designs based on this common principle can be calculated with Eq. 7·9. Since this research has been separately described, no attempt will be made to outline it here, beyond quoting certain results which may be useful in practical work.[1]

Figure 11·3 indicates the dependence of precision in temperature measurement on the length (L in Fig. 7·2) of immersion for the installation of Fig. 11·1. Scatter in readings, to be expected from using stock commercial materials, is also shown. This spread is among different installations made from the same spool of wire and operated with 350 to 650°F temperature difference between "hot" and "cold" junctions.[1]

Figure 11·4 shows the temperature dependence, found in a 7-hr test, of electrical-insulation resistance for the circumstances of the installations indicated in Figs. 11·1 and 11·7, respectively. For lifetimes in

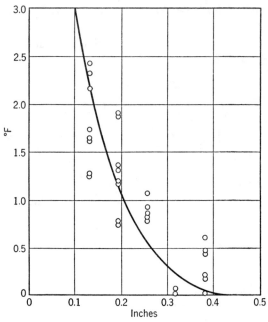

Fig. 11·3. Error, $t_2' - t_2$, °F vs. length of immersion, L, in.

service exceeding 7 hr, lower temperatures would be required for the same values of insulation resistance. The insulation resistance that will be required in a given job, as characterized by temperature patterns, precision requirements, etc., can be determined by reference to Fig. 5·15, Sec. 5·9.[1]

11·5 OTHER DESIGNS WITH LEADS EMERGING TOGETHER

Figure 11·5 shows an installation that differs from that of Fig. 11·1 in that the outer braid is not removed except for the portion of the wires made completely bare. To allow for the thickness of the outer braid, the hole that must be drilled in the body is correspondingly larger than for the installation shown in Fig. 11·1, i.e., No. 50 (0.070 in.) instead of No. 56 (0.046 in.). This greater diameter of the hole results in a decrease in the precision with which the point at which

Sec. 11·5 OTHER DESIGNS WITH LEADS EMERGING TOGETHER 147

the temperature is measured can be located. The presence of the outer braid on the wires at the point of their emergence from the hole adds substantially to the ruggedness of the installation. Unless conditions are severe, it is unnecessary to anchor the leads as shown in Fig. 11·2.[1]

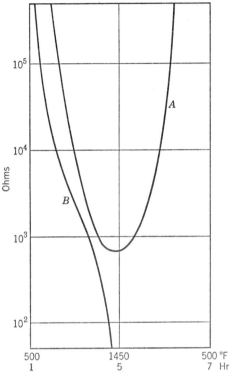

Fig. 11·4. Electrical-insulation resistance, ohms, vs. temperature, °F. *A*, Technical "B" Copper Cement insulation as in Fig. 11·1; *B*, G.E. No. 1201 Glyptal Red Enamel insulation as in Fig. 11·7.

Ordinarily, it suffices merely to daub the leads with thick G.E. 1201 Glyptal Red Enamel at the point of emergence.

Figure 11·6 shows a design that differs from that of Fig. 11·5 above in that No. 24 B & S gage (0.020 in.) glass-insulated, duplex, iron-against-constantan thermocouple wire is used. This size of wire requires that the hole in the body be No. 47 (0.078 in.) and that the bared portion be increased to ½ in. in length. A somewhat greater degree of ruggedness results. On the other hand, space requirements for the installation are larger, resulting in a decrease in the precision with which the location of the temperature actually measured is known.[1]

Figure 11·7 shows a design that can be resorted to where there is insufficient space to permit the installation shown in Fig. 11·1, or where

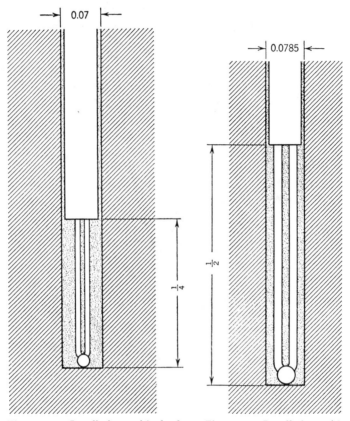

Fig. 11·5. Installation with leads emerging together.

Fig. 11·6. Installation with leads emerging together.

a higher degree of precision is required in the knowledge of the location of the temperature measured. A No. 65 (0.035 in.) hole is produced to the correct depth. Since the actual junction occurs at the point at which the two wires enter the bead, the hole must extend beyond the point at which it is desired to measure temperature, and by an amount equal to the length of the bead (see Fig. 5·5). It is usually advisable to decide on the bead length in advance and to trim the actual bead to this dimension by hand-grinding on fine emery cloth.

Sec. 11·5 OTHER DESIGNS WITH LEADS EMERGING TOGETHER 149

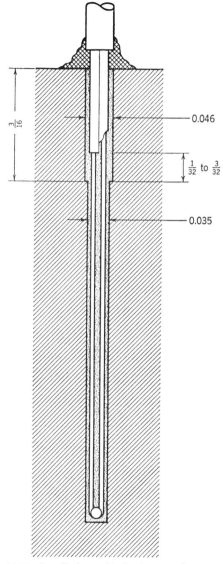

Fig. 11·7. Installation with leads emerging together.

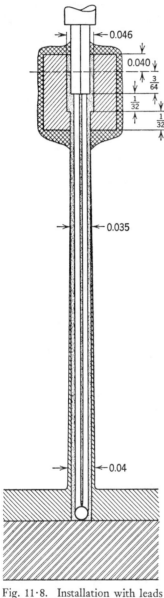

Fig. 11·8. Installation with leads emerging together, for application in a thin fin.

This hole is subsequently counterbored with a No. 56 drill (0.046 in.) to a depth of $\frac{3}{16}$ in.[1]

No. 30 B & S gage (0.010 in.) glass-insulated, duplex, iron-against-constantan thermocouple wire is used. The outer braid is removed for a length, after beading, of $\frac{1}{32}$ in. to $\frac{3}{64}$ in. greater than the depth of the hole. The wires are bared for a length, after beading, of $\frac{3}{32}$ in. to $\frac{5}{32}$ in. less than the depth of the hole. After beading and trimming of the bead to the proper length, the bare portions of the wires are spread just out of contact and coated with G.E. 1201 Glyptal Red Enamel. After air-drying, the lacquered ends are baked 15 min at 300°F. Because of the conspicuous color of this coating, complete coverage can be assured by careful inspection.

The junctions are cemented into the holes by the same procedure as for the Fig. 11·1 installation (see Sec. 11·2). The leads can then be anchored as indicated in Fig. 11·2. Thick G.E. 1201 Glyptal Red Enamel should be applied over the anchoring and at the point where the leads emerge from the hole. This coating should be permitted to become thoroughly dry before any rough handling occurs.

Because of the use of G.E. 1201 Glyptal Red Enamel baked at 300°F for electrical insulation, this design is limited in operating temperature to that at which the coating is baked, i.e., 300°F. Substitution of a suitable silicone paint for G.E. 1201 Glyptal Red Enamel will permit operating temperatures up to 500 or 1000°F. Coatings baked at temperatures higher than those recommended by the manufacturer tend to be more brittle,

Sec. 11·6 LEADS EMERGING SEPARATELY

and difficulties arise accordingly in the handling during installation.[1]

Figure 11·8 shows the preceding design slightly modified for installation in a thin fin. The fin shown tapers from 0.04-in. thickness at the base to $\frac{1}{32}$ in. at the tip. During the drilling of the No. 65 (0.035 in.) hole, it must be clamped between polished, hardened, tool-steel plates. A cylindrical brass fitting, drilled as shown and slotted to fit the fin, is driven on to reinforce the fin and support the leads at the point of their emergence. This fitting is cemented in position together with leads and the anchoring (see Fig. 11·2) by embedding in thick G.E. 1201 Glyptal Red Enamel.

11·6 LEADS EMERGING SEPARATELY

If the portion of the body in which the installation is to be made is not excessively thick, and if the side opposite the point of installation is accessible for work, the design shown in Fig. 11·9 can be employed. This installation can be performed in thinner sections than those of Secs. 11·2, 5. The surface on the opposite side of the body is disturbed by drilling and subsequent plugging, as indicated. The leads emerge separately on the same side of the body.[1]

The No. 76 (0.020 in.) holes are drilled at angles as shown in Fig. 11·9. The emerging hole is then counterbored by hand from the inside with a No. 60 (0.040 in.) drill to a depth of about $\frac{3}{16}$ in., after which the hole is tapped with a jeweler's tap to a depth of $\frac{1}{16}$ in.

The outer braid is removed from the No. 30 B & S gage (0.010 in.) glass-insulated, duplex, iron-against-constantan thermocouple wire for a length about $\frac{3}{4}$ in. greater than the wall thickness. The inner winding is removed for about $\frac{3}{8}$ in. at the end of each strand. The wires are threaded through the two holes to project an inch or more beyond each wall. The ends are twisted together and beaded (see Sec. 5·8). Lacquering with Technical "B" Copper Cement to provide electrical insulation is performed, as in the case of the installation shown in Fig. 11·1. After the lacquer has set, the wires are heavily painted with Technical "B" Copper Cement and moved in and out several times to "work" the cement up into the holes for half an inch or so. Then the wires are drawn in until the junction is flush. The plug shown in Fig. 11·9 is immediately driven or screwed in. Where operating conditions are such that mechanical forces and rubbing or abrasive action are absent from this inner or opposing surface, the plug can be omitted. Varnish-impregnated glass-braid "spaghetti" is slipped onto each lead. Anchoring is then performed as in Fig. 11·2.

The installation design shown in Fig. 11·10 represents the previous design modified so as to permit simpler drilling operations. A butt-welded junction (see Sec. 5·8) is used in contrast to the beaded junction employed in preceding designs. The junction can be placed at any point along the length of the wire. The characteristics of the Fig. 11·10 design in contrast to those of Fig. 11·9 result in the creation

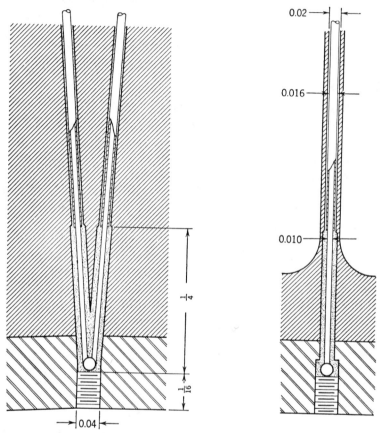

Fig. 11·9. Installation with leads emerging separately.

of a somewhat greater disturbance at the inner or opposing surface of the body.

Before installation, the insulation is stripped to the bare wire for ⅛ in. along each lead, and welding is performed. A marker, such as a collar or dab of red lacquer, is then applied at a suitable point and the distance measured (see Sec. 5·8) from the junction to this mark. The bare portion is coated with a thin mix of Technical "B" Copper Cement. After setting, this is varnished over with clear Glyptal. Technical "B"

Sec. 11·7 SMALL-SCALE INSTALLATIONS 153

Copper Cement is applied to the wires and "worked" into the holes. The wire is then threaded in one hole and out the other, to such a point that the marker is at a measured distance from the surface of

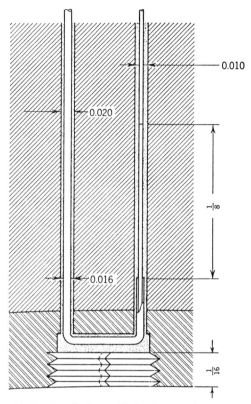

Fig. 11·10. Installation with leads emerging separately.

the work, thus locating the junction. The plugs can then be inserted. After the cement has set, the leads are anchored as in Fig. 11·2, except that a varnish-impregnated glass-braid "spaghetti" should first be slipped onto each wire.

11·7 SMALL-SCALE INSTALLATIONS

Figure 11·11 shows a design that can be used where space limitations are severe and temperature patterns complex, as, for example, in the electrodes of a spark plug. Installation can be effected by entering from one point on the surface of the body. The size of hole required is No. 73 (0.024 in.). The precision with which the location of the temperature measured is known is therefore correspondingly high. Once installed, such a design is rugged, and will withstand operating temperatures up to around 1000°F at the point the leads are anchored, and as high as 1400°F at the junction.[1]

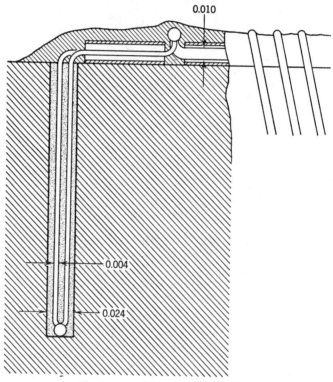

Fig. 11·11. Small-scale installation.

Number 38 B & S gage (0.004 in.) wires of platinum and platinum with 13 per cent rhodium are used for the thermocouple. The leads are made of No. 30 B & S gage (0.010-in.) wires, respectively, of the same materials. These materials can be duplicated very exactly if ordered "chemically pure." The lead wires are contained in individual tubes of braided glass-fiber "spaghetti." A suitable silicone paint or varnish is used to embed the leads and binding. Where operating temperatures permit, thick G.E. 1201 Glyptal Red Enamel can be substituted for silicone.[1]

Sec. 11·7 SMALL-SCALE INSTALLATIONS

The beaded junction is made in the usual manner (see Sec. 5·8). After beading, the 0.004-in. wires are carefully separated and coated with Technical "B" Copper Cement. After the junction has been cemented into the hole (see Sec. 11·2) with Technical "B" Copper Cement, the 0.010-in. extension leads are welded to the 0.004-in. wires. A short piece of "spaghetti" is threaded onto each wire. The ends are twisted together with the corresponding lead wires and beaded (see Sec. 5·8). The "spaghettis" on each lead wire and the short pieces on each junction wire are brought close to this point from either side. The two joints should be somewhat staggered, i.e., not exactly alongside each other. Before embedding in cement, care should be taken that the individual wires are well out of contact at points not surrounded by "spaghetti." The binding or anchoring is then completed.[1]

It may be necessary to install a junction at a location that cannot be reached by drilling a straight hole from a point on the surface of the body. For example, this is requisite when it is desired to install a junction in the tip of an outer electrode on a spark plug. In such circumstances, it is sometimes feasible to use an installation of the sort indicated in Fig. 11·12. For this installation, the operating temperature at the straight portion adjacent to the junction and underneath the plug can be as high as 1400°F. The temperature in the inclined hole through which the leads enter should not exceed 1000°F.

The No. 73 (0.024 in.) inclined hole is precision-drilled, as indicated

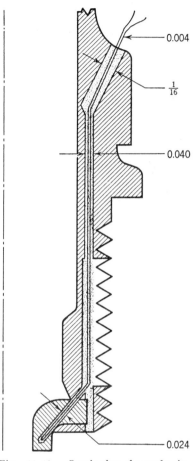

Fig. 11·12. Spark-plug-electrode installation.

on Fig. 11·12. Either the No. 52 ($\frac{1}{16}$-in.) inclined hole or the No. 60 (0.040-in.) vertical hole can be drilled next. The $\frac{1}{16}$-in.-wide slot shown is then milled out and the opening of the 0.040-in. hole plugged, if necessary, or left as indicated.

Number 38 B & S gage (0.004-in.) wires of platinum and platinum with 13 per cent rhodium are used. These are cut an inch or two longer than the developed length of the tortuous hole. After beading the junction (see Sec. 5·8), the wires forming the couple are stretched separated on a frame, cleaned with carbon tetrachloride, and coated with a very thin mixture of suitable silicone paint or varnish.

While remaining stretched on this frame, the couples are baked in air at around 650°F, or as recommended by the manufacturer, until the coating is dry and hard. The silicone varnish may not remain in a homogeneous layer but may tend to form into tiny beads surrounding the wire. The operation should be repeated, if necessary, until all portions of the wire are covered with such beads. The couples are then removed from the frame and the portions adjacent to the junction, i.e., the portions that ultimately enter the straight part of the hole, are coated with Technical "B" Copper Cement as for the installation shown in Fig. 11·1.[1]

Technical "B" Copper Cement is "worked" into the 0.024-in. hole, and the junction is introduced by way of the slot. After the cement has set, the leads are threaded back through the inclined holes to their normal point of emergence. When such fine wire is used, a suitably baked silicone coating is sufficiently flexible to permit bending, and yet hard enough to withstand the scraping involved in the operation of threading-through.[1]

The portions of the wires visible in the slot are carefully separated. The bottom of the slot is then filled with Technical "B" Copper Cement, and the cement permitted to set. In some applications it may be desirable to install a plug in the slot. No. 30 B & S gage (0.010-in.) leads are welded on and anchored as shown in Fig. 11·11. At the time of embedding the wires and binding in cement, an effort is made to work as much of this cement as possible into the hole around the wires in order to provide support for the wires at their point of emergence.[1]

11·8 LEADS EMERGING ON OPPOSITE SIDES OF THE BODY

Where conditions are such that the leads can be permitted to emerge on two opposing sides of the body (see Figs. 8·14, 16), a single straight hole can be utilized as indicated in Fig. 11·13. This type of design can be used instead of the arrangements shown on Figs. 11·9, 10. Each

Sec. 11·9 SHEATHING THE LEADS 157

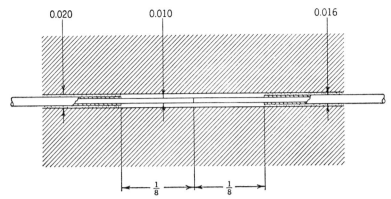

Fig. 11·13. Installation with leads emerging on opposite sides of body.

lead must then be anchored separately (see Fig. 11·2). It is advisable to slide a lacquer-impregnated glass-braid "spaghetti" around each lead and anchor these also.[1]

11·9 SHEATHING THE LEADS

For leads traversing the room from an installation above room temperature, insulation as furnished on the wire, or "spaghetti" made of glass braid impregnated with silicone, will serve most purposes. If the temperature of the leads nowhere exceeds 200°F, ordinary radio "spaghetti" is satisfactory.

If the leads are submerged in water, a highly waterproof tubing is needed. Rubber has been used (see Fig. 8·20). Suitable coating of fiber braids may suffice under certain circumstances.

Where the leads must pass through regions of corrosive fluids, liquids at elevated temperatures, abrasive streams, and other severe circumstances, sheathing in tubing of suitably corrosion-resistant metal is usually necessary (see Figs. 8·8, 11, 14, 15, 16; and Sec. 10·3). In order that it be effective as a metallic partition, all joints in such tubing, including that at the anchorage, must be autogenously welded, brazed, or soldered with corrosion-resistant solder, properly clamped and gasketed, or embedded in corrosion-resistant cement of sufficient temperature and stress resistance to suit the circumstances. Since the metal tubing affords no electrical insulation in itself, the wires must be adequately insulated within such tubing by separate means. For high-temperature service, two-hole ceramic "spaghetti" is used (see Fig. 8·8). For curving tubes, short sections or "beads" are employed. If the operating temperature is sufficiently low to permit and if corrosive fluids are present, this inner insulation should be of a sort resistant and impervious to such fluids.

For temperatures above those for which suitable metallic tubing is available, impervious ceramic tubing is used. Two-hole "ceramic-spaghetti" sections inside such tubing serve as electrical insulation. Clamped gaskets, cements, and seals of various sorts are used at the joints.

REFERENCE

1. H. D. Baker and E. A. Ryder, "A Method of Measuring Local Internal Temperatures in Solids," *Paper* 50-A-101, pp. 1–9, American Society of Mechanical Engineers, New York (1950).

12

TEMPERATURE GRADIENT INSTALLATION DESIGNS

12·1 TEMPERATURE GRADIENT AND HEAT FLOW

The rate of heat flow per unit of time by conduction at a point within a solid body is proportional to the thermal conductivity of the material of which the specified portion of the body is composed. It is also proportional to the steepness of the *temperature gradient* at this point in the direction of the heat flow. Temperature measurements at pairs of interior points can be utilized to determine temperature gradients along the lines joining the pairs of points, thereby affording data for computing the components of local heat-flow rate in these directions. Thus, the difference between the instantaneous values of the temperatures at two points divided by the distance between them equals the *mean value* of the component of temperature gradient in this interval of space.

The temperature pattern may be such that the gradient varies appreciably with location and differs from the mean at points within this interval. Such deviation from the mean will be less for smaller sizes of intervals. On the other hand, the absolute error in measurement of the distance between the two points is a larger fraction of this distance for small intervals, and the relative error (see Sec. 3·2) in measurement is correspondingly larger. A compromise must therefore be made. The length of interval must be chosen to suit each case according to the local complexity of the temperature field and the precision of the method to be used for ascertaining the distance between the two points for which the temperature difference is measured.

If the temperatures are measured separately and subtracted, the error in measurement enters the calculation twice, thus making the probable error larger than it would be if the difference were measured directly. Hence, for the measurement of gradients, the two junctions of the thermocouple circuit are placed, respectively, at the two end-points of this interval. Readings then correspond to the difference between the temperatures at these two points.

When the direction of the heat flow is obvious or known by independent means, the magnitude of the temperature gradient can be determined by measuring the difference between temperatures at two points located along this direction. In general, however, this direction is not known, and to measure the resultant temperature gradient at a given point, it is necessary to measure the local components of tempera-

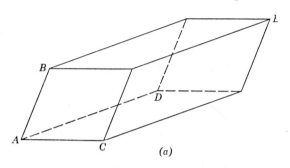

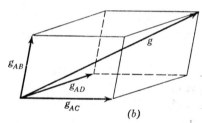

Fig. 12·1. Temperature gradient. *a*, space diagram; *b*, vector diagram.

ture gradient in three directions orthogonal or obliquely inclined to one another. Thus, referring to Fig. 12·1, let t_A be the temperature, °F, at point A; t_B be the temperature, °F, at point B; t_C be the temperature, °F, at point C; and t_D be the temperature, °F, at point D. Let g_{AB} be the mean value of temperature gradient in the direction of A to B over the interval AB, °F/ft; g_{AC} be this gradient for interval AC; and g_{AD} be this gradient for interval AD. Let L_{AB} be the distance between points A and B, ft; L_{AC} be this distance for points A and C; and L_{AD} be this distance for points A and D. Then

$$g_{AB} = (t_B - t_A)/L_{AB} \tag{12·1}$$

$$g_{AC} = (t_C - t_A)/L_{AC} \tag{12·2}$$

$$g_{AD} = (t_D - t_A)/L_{AD} \tag{12·3}$$

Sec. 12·2   LEADS EMERGING TOGETHER   161

The resultant temperature gradient g is in the direction of, and proportional to, the length of the diagonal of the parallelepiped (see b in Fig. 12·1). The resultant g is determined by graphical or trigonometric means.[1]

12·2 TEMPERATURE GRADIENT AT AN INTERIOR POINT—LEADS EMERGING TOGETHER

Figure 12·2 indicates the essential features of a convenient design for the measurement of temperature gradient at a point. It can be installed in any solid material, and requires approach from only one direction. The depth of insertion in the material may range from $\%_{16}$ in. to 6 in. or more. In the completed form it has sufficient ruggedness to serve average industrial test-stand requirements and can be produced from commercially available materials. Its precision in the measurement of temperature difference can be to within 0.25 per cent to 0.75 per cent (or less for special wire), if adequate care is taken to reduce parasitic thermoelectric and voltaic emfs, and depending on the sensitivity of the indicating instruments.[2]

The distance between the two junctions, labeled $3/_{16}$ in. on Fig. 12·2, can be varied to suit the circumstances. This interval can be located anywhere along the length of the wire, provided that the outermost junction is not closer than $3/_{16}$ in. to the surface of the body. The precision of the determination of the distance between the two points for which temperature difference is measured is usually to within $1/_{64}$ in. Installations of this design can be used at temperatures up to around 1400°F and at lower

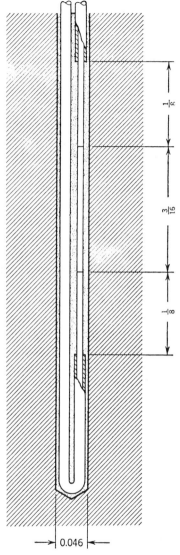

Fig. 12·2. Gradient installation with leads emerging together.

temperatures down to about −425°F. The design, as described below, is not intended for highly humid or submarine conditions. Suitable coating or sheathing of the leads will, however, adapt it to such service.[2]

The installation is performed as follows. Number 30 B & S gage (0.010-in.) glass-insulated, duplex, iron-against-constantan thermocouple wire is used. The outer braid is removed from the entire length of wire used. The constantan strand is divided into two parts, each of which must be long enough to reach to the auxiliary instrumentation. The insulation is stripped to the bare wire on each section of the constantan for $\frac{1}{8}$ in. from the end. These wires are then joined through a $\frac{3}{16}$-in. length, cut from the strand of iron wire, by butt-welding (see Sec. 5·8). The distance between the two junctions is measured. The longer wire is then doubled back as indicated in Fig. 12·2 and a marker, such as a collar or dab of red lacquer, is applied at a suitable point. The distance from one of the junctions to this mark is measured. The bare portion is coated with a thin mix of Technical "B" Copper Cement. After setting, this is varnished with clear Glyptal (see Sec. 10·2). A No. 56 (0.046 in.) hole is drilled to the required depth in the material at the correct location and angle. The hole is cleaned with carbon tetrachloride. Technical "B" Copper Cement is mixed to a cream and the hole is filled with this cream. By using a wire about half the diameter of the hole as a "piston," it is quite easy to "work" (or "pump") the cement to the bottom of the hole and be sure of having it there. When the hole will absorb no more cement, the prepared junction is painted with the copper-cement cream and immediately inserted into the bottom of the hole. It is lifted up and down a few times and then held at the proper position, as determined by the distance from the marker to the surface of the body, until the cement has set. This will require about 15 min. The progress of the setting process can be observed by watching the gradual thickening of the portion of cement left over.

When the cement has set, the installation is completed except for means which may be employed to protect the leads. A suitable size of varnish-impregnated glass-braid "spaghetti" should be threaded over the leads up to the point at which they emerge from the body material. Then anchoring is performed (see Sec. 11·3).

Figure 12·3 shows a modification of the design of Fig. 12·2. This modified design occupies less space and can be expected to provide a somewhat more precise determination of the distance between the two points for which temperature difference is measured.

A No. 65 (0.035 in.) hole is produced to the proper depth. This

Sec. 12·3   LEADS EMERGING SEPARATELY

hole is subsequently counterbored to No. 56 (0.46 in.) for a depth of ³⁄₁₆ in.

No. 30 B & S gage (0.010 in.) glass-insulated, duplex, iron-against-constantan thermocouple wire is used. The wire is bared for $\frac{1}{32}$ in. to $\frac{3}{32}$ in. beyond that section of the wire which will ultimately remain within the 0.035-in. portion of the hole. After the junctions have been made, the distance between them measured, and the marker applied (as for Fig. 12·2), the bare portions of the wires are spread just out of contact and coated with G.E. 1201 Glyptal Red Enamel. After air-drying, the coated end-section of each wire is baked 15 min at 300°F. The conspicuous color of this coating permits assurance, by careful inspection, of complete coverage.

The couple is cemented into the hole by the same procedure as that described for the installation shown in Fig. 12·2. Lacquer-impregnated glass-braid "spaghetti" is slipped over the leads and anchoring is performed.

Because of the use, for electrical insulation, of G.E. 1201 Glyptal Red Enamel baked at 300°F, this installation is limited in operating temperature to 300°F. Substitution of a suitable silicone varnish or paint for G.E. 1201 Glyptal Red Enamel will permit operating temperatures up to 500 or 1000°F. For use in thin webs or fins, this design can be modified to include a supporting bushing [2] (see Fig. 11·8).

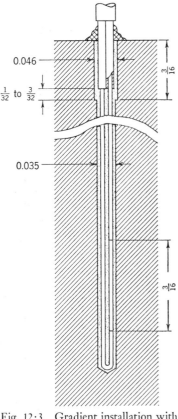

Fig. 12·3. Gradient installation with leads emerging together.

12·3 TEMPERATURE GRADIENT AT AN INTERIOR POINT— LEADS EMERGING SEPARATELY

Figure 12·4 shows a design permitting installation in a thinner web of material than that required for Fig. 12·3. Greater precision can be expected in ascertaining the distance between the two points for which temperature difference is measured. Installations made according to this design can be used at temperatures up to around 1400°F and at lower temperatures down to about −425°F.[2]

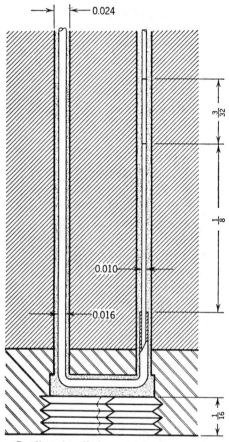

Fig. 12·4. Gradient installation with leads emerging separately.

Two No. 73 (0.024 in.) holes are drilled as shown. No. 30 B & S gage (0.010-in.) glass-insulated, duplex, iron-against-constantan thermocouple wire is used. The outer braid is removed from the entire length of wire. The constantan strand is divided into two parts, each of which must be long enough to reach to the auxiliary instrumentation. The insulation is stripped to the bare wire for $\frac{1}{8}$ in. from the end on each constantan strip. The bared tips are then joined through a $\frac{3}{32}$-in. length and cut from the strand of iron wire by butt-welding. The distance between these two junctions is then measured (see Sec. 5·8).

A marker, such as a collar or dab of red lacquer, is applied at a suitable point and the distance from one of the junctions to this mark

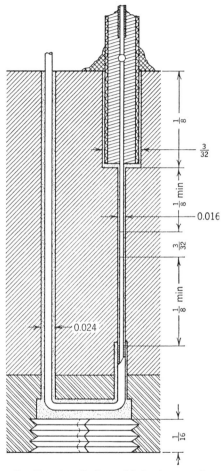

Fig. 12·5. Gradient installation with leads emerging separately.

is measured. The bare portion is coated with a thin mix of Technical "B" Copper Cement. After setting, this is varnished with clear Glyptal. Technical "B" Copper Cement is applied to the wires and worked into the holes. The wire is threaded in one hole and out the other, stopping at such a point that the marker is at a measured distance from the surface of the work, thus locating the pair of junctions. The plugs are then inserted. After the cement has set, a varnish-impregnated glass-braid "spaghetti" is slipped on over each lead. The leads are then anchored.

Figure 12·5 shows a modification of the preceding design, intended to provide greater precision in the measurement of temperature gradient. Thus, the portion of the hole in which the junctions are located is smaller, i.e., No. 78 (0.016 in.).

The procedure in drilling is as follows. Drill the No. 73 (0.024-in.) hole through. Then drill the No. 42 ($3/32$-in.) counterbore $1/8$ in. deep. Drill the No. 78 (0.016-in.) hole through from the lower side (see Fig. 12·5) to this $3/32$-in. counterbore. Counterbore to No. 73 (0.024 in.) as shown. Drill and tap for the plugs indicated.

Number 30 B & S gage (0.010-in.) glass-insulated, duplex, iron-against-constantan thermocouple wire is used. The outer braid is removed from the entire length of wire. The constantan strand is divided into two parts, each of which must be long enough to reach to the auxiliary instrumentation. The insulation is stripped to the bare wire for $1/8$ in. from the end of one constantan strip and for $5/8$ in. on the other. The bare ends are then joined through a $3/32$-in. length cut from the strand of iron wire by butt-welding. The distance between these two junctions is measured by the procedure explained in Sec. 5·8. The constantan lead with the larger bared portion is severed $3/16$ in. from the beginning of the insulation, i.e., $7/16$ in. from the nearest junction. The bare portion that is to enter the hole is then coated with G.E. 1201 Glyptal Red Enamel except for a $3/16$-in. length at the end, and baked for 15 min at 300°F. Technical "B" Copper Cement is applied to the lead carrying the junctions and is "worked" into the holes. This lead is then threaded in the 0.024-in. diameter hole and back out the 0.016-in. diameter hole, stopping at such a point that the bare end is at a measured distance from the surface of the work, thus locating the pair of junctions. The plugs are then inserted. After the cement has set, the constantan lead (severed above) is rejoined by twisting together a $1/8$-in. length of each strand and beading. The bead and adjacent bare portion are then coated with G.E. 1201 Glyptal Red Enamel. Varnish-impregnated glass-braid "spaghetti" is then slipped on over each lead. One of these is cemented into the $3/32$-in. counterbore with thick, G.E. 1201 Glyptal Red Enamel. The leads are then anchored.

12·4 TEMPERATURE GRADIENT AT AN INTERIOR POINT—LEADS EMERGING ON OPPOSITE SIDES OF THE BODY

Where conditions are such that the leads can be permitted to emerge on two opposing sides of the body, a single straight hole can be used instead of the arrangements shown on Figs. 12·4, 5 above. Each lead must then be anchored separately [2] (see Fig. 11·13).

REFERENCES

1. F. B. Seely and N. E. Ensign, *Analytical Mechanics for Engineers*, pp. 35–37, 53–55, John Wiley & Sons, New York, 1941.
2. H. D. Baker and E. A. Ryder, "Method of Measuring Local Internal Temperatures in Solids," *Paper* 50-A-101, pp. 1–9, American Society of Mechanical Engineers, New York (1950).

13

CONCLUSION TO VOLUME I

13·1 DESIGNS

Various thermocouple designs have been indicated in Chs. 8, 11 and 12 of this volume. These are intended for specific classes of problems in the measurement of internal temperatures and temperature gradients in solid bodies. Each design has a certain range of applicability; it is not to be supposed, however, that all possible types of problems have been covered. The purpose of Chs. 11 and 12 has been to suggest possible designs and to indicate the features that are essential to the cemented type of installation. Various modifications of the specified designs will suggest themselves for other situations.

13·2 BASIC INFORMATION

Chapters 3, 4, 5, 7, and 9 supply basic information for the execution of the specific designs illustrated in Chs. 8, 11, and 12. This basic information will also be useful for solving special problems that may arise in particular circumstances.

13·3 APPLICATION

In applying the designs, it must always be remembered that the purpose of each of these designs is to establish and maintain in service the thermoelectric junction in the best possible thermal communication with the parent metal, at the same time creating minimum disturbance in the temperature field within the body.

Volume II deals with problems in the measurement of very low and very high temperatures; surface temperatures; rapidly changing temperatures and temperatures in rapidly moving bodies; temperatures of liquids, gases, and flames; and temperatures of gases with entrained particles or coexistence of states.

NAME INDEX

Agnew, W. G., 22
Akin, G. A., 90, 92, 103
American Gas Association, 21
American Society for Testing Materials, 12, 21
American Society of Mechanical Engineers, 12, 20, 21, 61, 62, 67, 127, 139
Ames Co., W. V-B, 53, 139
Arreger, C. E., 16
Ashman, A. O., 102
Aston, J. G., 61

Bachman, C. H., 140
Badger, W. L., 92, 104
Bailey, N. P., 63, 105
Baimakoff, Y. V., 63
Baker, E. M., 103
Baker, H. D., 12, 67, 83, 106, 158, 167
Baker, H. W., 103-105
Barkley, J. F., 102
Barr, W. E., 140
Bass, E. L., 104
Bear, H. R., 140
Becker, J. A., 20
Bell, E. R., 102
Bennett, H., 139
Bernardo, E., 102
Biddle Co., James G., 140
Blake Co., Edward, 113
Boelter, L. M. K., 106
Bogdan, L. J., 105
Bole, G. A., 21
Bonilla, C. F., 104
Brenner, B., 62
Bride, W. L., 105
British Standards Institution, 20, 67
Brown, G. S., 67

Brown Instrument Co., 20
Bryant, E. M., 105
Busse, J., 12
Buttner, H. J., 22
Byerly, W. E., 83

Caldwell, F. R., 140
Campbell, D. P., 67
Carnot, 2, 3
Carter, H. J., 63
Cartwright, C. H., 139
Cataldo, J. T., 63
Catlin, A. A., 105
Celsius, 4, 5
Centinkale, T. N., 83
Central Scientific Co., 138
Charles, 1
Chelko, L. J., 105
Chestnut, H., 67
Clark, F. H., 140
Clark, W., 22
Claypoole, W., 67
Clement, J. K., 95, 104
Colburn, A. P., 95, 104
Comstock, G. C., 29
Corey, R. C., 21
Corrington, L. C., 103
Corrucini, R. J., 61
Craig, D. N., 21
Curie, 9
Curtis, C. E., 140

Dahl, A. I., 61-63
Danforth, W. E., 62
Davidson, P., 61
Dewar, 7, 59, 60
Dickson, G., 140

NAME INDEX

Dike, K. C., 140
Dudugjian, C., 103
Dunlap, M. E., 102
DuPont, E. I. de Nemours & Co., 134
Durham, J. G., 12

Eckelt, O., 126
Eckman, D. P., 67
Eichelberg, G., 105
Emmons, H., 63
Engineering Societies Library, 62
Ensign, N. E., 167

Fahrenheit, 1, 5
Fairchild, C. O., 12, 63
Farrington, G. H., 67
Feagon, R. A., Jr., 104
Féry, 19
Fiock, E. F., 63
Fishenden, M., 83
Fitterer, G. R., 62
Foote, P. D., 12, 63
Forsythe, W. E., 21
French, L. G., 126
Fuller, D. D., 67

Gangler, J. J., 140
Garland, C. M., 95, 104
Garrison, J. B., 20
Gaydon, A. G., 22
Geffner, J., 29
General Electric Co., 129
Gibson, A. H., 91, 103
Gilbert, W. W., 126
Goodrich, C. L., 126
Gordon, C. L., 140
Gowens, G. J., 62
Green, C. B., 20
Guilbert, A. G., 106
Guthmann, K., 22

Hall, E. L., 140
Halliday, D., 20
Hamilton Tool Co., 119
Hammel, E. F., 63
Hardy, J. D., 103
Harrison, T. R., 12, 63
Harrison, W. N., 140
Hase, R., 22

Hawkins, G. A., 83
Hayward, R., 139
Hebbard, G. M., 92, 104
Herzfeld, K. F., 12
Hidnert, P., 140
Hiergesell, V., 12
Hoagland, F. O., 126
Hoge, H. J., 12, 140
Homewood, C. F., 62
Hougen, O. A., 95, 104
Hug, K., 98, 105
Humphreys, C. G. R., 102
Hyman, S. C., 104

Jakob, M., 21, 83, 105
Judge, A. W., 104, 105

Kaasa, O. G., 102
Kacena, F. B., 102
Kaufman, A. B., 62
Kaufman, S. J., 102, 105
Kelvin, 4
Keyes, F. G., 11
Keyser, P. V., 104, 105
Kinkul'kin, B., 62
Kleinberg, R. M., 102
Kleinhans, F. B., 126
Kline, G. M., 141
Kreisinger, H., 102
Krembs and Co., 49
Kurbaum, 19

Laserson, G. L., 12
Lauritzen, J. I., 61
Lawson, A. W., 20
Lawson, E. C., Jr., 102
Leeds & Northrup Co., 103
Lenin & Son, Louis, 119
Lennie, A. M., 126
Lewis, A. B., 140
Lewis, B., 22
Lieneweg, F., 106
Lindner, C. T., 12
Liquid Steel Subcommittee, 62
Liston, M. D., 67
Long, R. A., 140

MacCoull, N., 103
McAdams, W. H., 83, 90, 92, 103

NAME INDEX

McNutt, J. E., 140

Manganiello, E. J., 102
Mayer, R. W., 67
Mebs, R. W., 63
Micro Products Co., 53
Miller, E. F., 104, 105
Miller, M. A., 106
Modes, E. E., 20
Mohler, F. L., 12
Mohun, W. A., 92, 104, 106
Moore, D. G., 140
Morgan, F. H., 62
Mucklow, G. F., 105
Mueller, A. C., 103
Mulcahy, B. A., 102
Murrell, T. A., 11

Nägel, A., 98, 105
Naeser, G., 21
Nail, N. R., 22
National Bureau of Standards, 9–12, 45, 46, 62
National Jet Co., 119
Neher, H. V., 139
Norton, E. W., 126

Office of Technical Services, 139
Ohm, 61, 65
Orton, Edward Jr., Ceramic Foundation, 22

Patton, E. L., 104
Pearlman, D., 22
Pearson, G. L., 20
Peters, M. D., 103
Peterson, W. S., 92, 104, 106
Pinkel, B., 102
Pohl, H. A., 21
Povolny, J. H., 105
Pumphrey, W. I., 62
Purdy, J. F., 87, 88, 102, 106

Quigley, H. C., 62
Quinn, C. E., 67

Rankine, 5
Reid, W. T., 21
Reiher, H., 95, 104
Rendel, T. B., 103

Richmond, J. C., 140
Ridgway, R. R., 62
Riehm, W., 104
Rinker, R. C., 141
Robards, C. F., 140
Roebuck, J. R., 11
Roeser, W. F., 12, 47, 61–63
Romie, F. E., 106
Rosenthal, I., 22
Royds, R., 102, 105
Ryder, E. A., 83, 103, 105, 106, 158, 167

Sanders, J. C., 103
Sargeant, W. E., 67
Sauereisen, C. F., 140
Saylor, C. P., 89, 103
Scheele, J. M. B., 126
Schlecht, W. G., 140
Schlieren, 20
Schoonover, I. C., 141
Schueler, L. B., 102
Schultz, A., 21, 61, 62
Scott, G. G., 67
Seely, F. B., 167
Shenker, H., 61
Shoemaker, F. G., 102
Smith, I. B., 105
Smith, L. F., 106
Souder, W., 141
Spear, E. B., 87, 88, 102, 106
Squire, C. F., 12
Steven, G., 62
Stimson, H. F., 7, 12
Stoll, A. M., 103
Strong, J., 139
Such, I. H., 126
Suits, C. G., 22
Sutor, A. T., 103
Swanger, W. H., 140

Taylor Manufacturing Co., 119
Teetor, M. O., 105
Teletronics Laboratory, 119
Trott, W. J., 63
Troy, W. C., 62
Tyte, L. C., 22

Uhlig, H. H., 140
Underwood, A. F., 105

United States Department of Commerce, 21
United States Treasury Department, 21
Urback, F., 22

Valerino, M. F., 102, 105
von Elbe, G., 22

Wallace, M. W., 22
Warner, D. K., 61
Watson, F. R. B., 104
Weills, N. D., 83, 103, 105

Weller, C. T., 62
Wendt, L. A., 103
Wensel, H. T., 12, 47, 62
White, W. P., 29, 67
Whitford, A. E., 139
Wichers, E., 140
Wile, D. D., 62
Wilsted, H. D., 103
Wood, W. D., 12
Worthing, A. G., 20, 29

Zipkin, M. A., 103

SUBJECT INDEX

Accidental errors, 25–27 (*see also* Errors)
Acids, 129, 132, 134, 135
Adjustable angle-tilting table, 118, 119
Alkalies, 129, 134, 135
Alkyd resin, 129 (*see also* Glyptal)
Alpha particles, 19
Alumel, 40, 41, 49
Alumina, 124, 125, 132, 135, 136
Aluminum, 109, 126
Aluminum oxide, 124, 125, 132, 135, 136
Amber, 134
Ambient temperature, 65, 68, 70–74, 128, 145–147, 157, 158
Amplifiers, electronic, 66, 67
Antimony, 7
Arkansas stone, hard, 113
Asbestos, 91, 130, 134

Bakelite, 92, 94
Baths, thermostated, 47
Beeswax, 129, 138 (*see also* Coatings)
Benzoic acid cell, 11 (*see also* Temperature scale, fixed points for)
Bevel protractor, 119
Bimetallic strip, 16 (*see also* Thermometers)
Boiling point, 4, 6, 11 (*see also* Temperature scale, fixed points for)
Brass, 92, 93, 109, 126, 131, 132
Brazing, 95–98, 124, 125, 133, 137 (*see also* Soldering)
Brittleness, from contamination, 41, 42, 49, 51, 131–133 (*see also* Contamination)
Bronze, 109, 126, 132

Calculation technique for design, 68–83, 100, 101
Calibration, for International Temperature Scale, 6, 7 (*see also* Temperature scale)
of thermocouples, 6, 7, 44–46, 100, 101 (*see also* Wire, thermocouple)
of thermometers, mercury-in-glass, 10, 11 (*see also* Thermometers)
Capacity, thermal, 68
Carbide, cemented, 124, 125, 133
Carbon, 41, 42, 132, 135, 137
Carbon tetrachloride, 122, 144, 156, 162
Carnot cycle, 2, 3
Celsius scale, 4, 5 (*see also* Temperature scale)
Cement, air-drying, 138, 139
chemically setting, 138, 139
melting, 129, 138
porcelain, 101, 132
Portland, 124
Technical "B" Copper, 139, 144–147, 150–153, 156, 162, 165, 166
thermosetting, 135–139
Cemented installations, 76–83, 94–96, 142–166 (*see also* Junctions; Thermocouple thermometer)
Ceramics, 89, 90, 125, 135–137, 157–158
Ceresin wax, 39, 129, 134 (*see also* Coatings)
Charles' law, 1, 2
Chromel P, 40–42, 86
Chromium, 132
Circuits, multiple-couple, 58, 65
thermocouple, junctions, 47–54 (*see also* Junctions)

173

SUBJECT INDEX

Circuits, thermocouple, laws of, 37
   multiple-metal, 35–37, 58, 59
   simple, 34, 35
   splices, 55, 154, 155
   switches, 37, 58, 66
  thermopile, 59, 65
Classes of thermometers, 14 (see also Thermometers)
Coatings, moisture-repellent, 39, 129–131, 134, 143, 157, 158, 162
  protective, 93, 128–131, 143, 145–147, 157, 158, 162
   ceramic, 133
   enamels, 39, 60, 92, 130, 144–147, 150–152, 154, 163, 166
   Glyptal, 96, 129, 144–147, 150–152, 154, 163, 165, 166
   paints, 130, 156
   silicone, 96, 130, 134, 135, 139, 156, 157
   spar varnish, 129, 130
   waxes, 39, 129, 134, 138
Cobalt, 124
Color indicators, 18, 19
Compound table, 118, 119
Compound vise, 119, 120
Conditions, characteristic, 31, 32, 128, 153, 159–161
Conductance, fluid-boundary, 72–83
  radiation-boundary, 72–83
  surface-boundary, 71–83
  thermal, of solids, 68–83, 128
Constantan, 39–43, 86, 88, 90, 93, 96, 144–147, 150, 151, 162–164, 166 (see also Wire, thermocouple)
Contact, electrical, 48
  thermal, 68, 70–83, 98, 99, 142, 145, 146
Contamination, brittleness from, 41, 42, 49, 51, 131–133
  inhomogeneity from, parasitic emfs, 36–38, 41, 43–45, 48, 49, 90, 131–133 (see also Parasitic emfs)
Control, 64
Copper, 38, 39, 60, 88, 93, 94, 109, 126, 132 (see also Wire, thermocouple)

Corrections, thermocouple calibration, 44–46, 100, 101 (see also Wire, thermocouple)
Corundum, 135
Cotton, 130, 135
Countersinking tool, 121
Crayons, temperature-indicating, 18, 19
Cryostat, 47
Curie law, 9

Data, treatment of, 26 (see also Errors)
De Khotinsky cement, 138
Design, apparatus, components of, 31
  principles and techniques of, 23, 30–32, 68–83, 100, 101, 128
  indicating-instrumentation, 64–67 (see also Instruments)
  thermocouple installation, cemented types, 76–83, 142–166 (see also Junctions; Thermocouple thermometer)
  survey of types, 84–101
Dewar flask, 7, 59, 60
Diamond grit, 124, 125
Drift in drilling, 108, 114, 123, 124, 126 (see also Drills)
Drills, broken, removal of, 117, 118, 126
  chip removal from, 114, 115, 122, 123
  chisel edge of, 110, 121
  chisel-point, 110
  cutting fluids for, 115, 116, 125
  depth of hole, measurement of, 108, 123, 124
  diamond-point, 110
  extended-shank, 110–112
  fishtail, 110
  fluted length of, 110, 114–116
  lip-relief angle of, 110, 113, 114
  locating starting holes for, 108, 109, 119–124
  machines for, 116–119
  making of special, 111–114
  pivot-point, 109, 110, 124
  point angle of, 110, 114
  runout of, 108, 114, 123, 124, 126
  shanks of, 110–112
  sharpening of, 112–114
  small, use of, 114–116
  speeds for, 114, 117, 121

SUBJECT INDEX

Drills, Swiss-made, 109
  twist, nomenclature of, 109, 110
  use of, in hard and abrasive materials, 124–126
  with diamond cutting points embedded in tips, 124

Electric-resistance thermometers, 6, 15, 16 (see also Thermometers)
Electrical insulation, 55–58, 72–74, 128–131, 134–136, 143, 145–147, 150, 162, 163 (see also Coatings; Sheathing)
Electrical leakage, 39, 55–57, 128, 145–147, 157, 158, 162 (see also Insulation, electrical)
Electrochemical method, broken drill removal by, 126 (see also Drills)
Emery, 125, 144
Enamel, 39, 60, 92, 130, 144–147, 150–152, 154, 163, 166 (see also Coatings)
Errors, absolute, 24, 81–83, 159
  accidental, treatment of, 25–27
  accumulation of, total, 28, 81–83
  allowable, 25, 28, 81, 83
  calculation of, 68–83, 100–101
  individual, sources of, 28, 68–83, 90, 95, 96, 108, 109, 159
  location, 68, 108, 109, 159
  probable, 25, 28, 81–83
  relative, 24, 159
  systematic, treatment of, 25, 27, 28, 81–83

Fahrenheit scale, 1, 5 (see also Temperature scale)
Fairing of data, 26 (see also Errors)
Fixed points, 1, 5–7, 10, 11, 45, 59, 60 (see also Temperature scale)
Flux, soldering, 39, 48, 49, 124, 125, 133 (see also Soldering)
Furnaces, for thermocouple calibration, 46 (see also Calibration)

Galvanometer, 42, 65–67, 71 (see also Instruments)
Gaskets, 91, 99, 131, 134, 157, 158
Gasoline, 129
Glass, 97, 124, 125, 130, 136

Glass fiber, 81, 86, 94, 130, 134, 135, 143, 150, 151, 153, 157, 162–164, 166
Glyptal, 96, 129, 144–147, 150–152, 154, 163, 165, 166 (see also Coatings)
Gold, 7, 43, 132
Gradient, temperature, 76–83, 142, 159–167

Heat transfer, 68–83, 128, 145, 146, 159–161
Heresite lacquer, 93 (see also Coatings)
High-speed steel, hardening of, 111
Holes, drilling requirements for, 70, 84–87, 107–109, 143, 153–156
Homogeneity, 36–38, 43–45, 48, 49, 90, 131–133 (see also Wire, thermocouple)
Hydrogen, 41
Hypodermic needle, 123
Hypsometer, 11

Ice point, 1, 4, 6, 7, 10, 11, 45, 59, 60 (see also Temperature scale, fixed points for)
Immersion, required length of, 72, 76–83, 142, 145, 146, 148
Inconel, 132
Instruments, indicating, 31, 37, 38, 43, 64–67, 71, 82, 83
  primary-standard, 3, 9
  standards-laboratories certification of, 5
  temperature measurement, survey of methods for, 13–21
  working-standard, 9–11
Insulation, electrical, 55–58, 72–74, 128–131, 134–136, 143, 145–147, 150, 162, 163 (see also Coatings; Sheathing)
  tests, 145–147
  thermal (see Conductance)
Interferometer, temperature measurement by, 20
Invar, 132
Iridium, 42 (see also Wire, thermocouple)
Iron, 40, 94, 109, 132, 143–147, 150, 151, 162–164, 166 (see also Wire, thermocouple)

SUBJECT INDEX

Jewels, 125
Johnson noise, 13, 65, 67
Julius suspension, 66 (see also Vibrations)
Junction box, 58, 71 (see also Circuits, thermocouple)
Junctions, thermocouple, 34, 47–54 (see also Circuits; Wire; Thermocouple thermometer)
   automobile-tire, 86–89
   beaded, 84, 85, 98
   biological-tissue, 89
   broken, welded, 54, 90
   butt-welded, 52–54, 94, 95, 166
   cemented, 76–83, 94–96, 142–166
   electroplated production of, 54
   liquid-contact, 91, 92
   peened, 84–87
   pierced-in, 87–89, 107
   plugs with, 96–100
   pressed-in-plug, 99–101
   screw-plug, 90, 91, 96–98
   soldered, 92–96, 97
   staked-in, 99–101
   surface-groove, 95, 96
   taper-plug, 98–100
   tube-wall, 84, 85, 92–96
   welded, 89, 90

Kelvin scale, 4 (see also Temperature scale)
Kerosene, 60, 113, 116

Lacquers (see Coatings)
Lag, 69
Lapping, 125, 126
Lead, 132
Leads, thermocouple (see Thermocouple thermometer, leads for)
Leakage, electrical, 39, 55–57, 128, 145–147, 157, 158, 162 (see also Insulation, electrical)
Least squares, method of, 26 (see also Errors)
Leather, 134
Line-reversal method, Féry, 19
Luminescent phosphors, 19

Magnesia, 86, 137
Magnesium oxide, 86, 137

Manganin, 38
Measurements, precision of, 25–28, 47, 67–69, 71, 95, 100, 101, 108, 109, 143, 145, 146, 148, 159–161 (see also Errors)
Mercury, 91, 92
Mica, 41, 132, 134, 136, 137
Micrometer caliper, 124
Moisture-repellent coatings, 39, 130, 131, 134, 143, 157, 158, 162 (see also Coatings)
Molybdenum, 42, 124, 133
Monel, 132
Mullite, 136

Nichrome, 132
Nickel, 94, 95, 132
Nonequilibrium conditions, occurrence of, 2, 3, 19, 69, 70, 159–161

Ohm's law, 61, 65
Olive oil, 125
Orifice, method of, 20
Oxygen, 6, 131–134 (see also Temperature scale, fixed points for)

Paints, protective, 130, 156 (see also Coatings)
   temperature-indicating, 18, 19
Paraffin wax, 39, 129 (see also Coatings)
Parallel, thermocouples in, 59 (see also Circuits)
Parasitic emfs, thermoelectric, 37, 38, 41, 48, 61, 65, 66, 81, 90, 145, 146
   voltaic, 38, 39, 55–58, 61, 65, 66, 81, 145, 146
Phosphorus, **19**, **41**
Photography, infrared, 19
   Schlieren, 20
   with luminescent phosphors, 19
Pivot-point drills, 109, 110 (see also Drills)
Plastics, 109, 130, 134, 135, 137
Platinum, 7, 40, 41, 49, 131, 132, 154, 155 (see also Wire, thermocouple)
Platinum-resistance thermometer, 6 (see also Thermometers)
Polytetrafluoroethylene, 134

SUBJECT INDEX

Porcelain, 41, 91, 97, 124, 132, 135, 136
Potentiometer, 66, 71
Precision, 23, 25, 47, 58, 61, 68, 95, 100, 101, 143, 145, 148, 159–161 (see also Errors)
Protection tubes (see Sheathing)
Protective coatings (see Coatings, protective)
Pseudotemperature scale, 4
Pyrex, 92
Pyrometer, 7, 13, 14 (see also Thermometers)
   optical, 17, 18
   radiation, 14, 17, 18, 46
   two-color, 17, 18
Pyrometric cones, 18

Quartz, 100

Radiation, blackbody, 7, 19 (see also Pyrometer)
   boundary conductance due to, 72–83 (see also Conductance)
   temperature measurement by, 7, 13, 14, 17–19, 46 (see also Pyrometer)
Rankine scale, 4, 5 (see also Temperature scale)
Reamer, square-end, 122, 124 (see also Drills)
Recording, 64
Refractories, nonmetallic, 89, 90, 125, 135–137, 157, 158
Repetition of readings, 26, 27 (see also Errors)
Resin, cured, 92, 93, 129
Resistance, electric, 6, 55–58, 61, 65 (see also Insulation)
   electrical contact, 48
   thermal, 68–83, 128 (see also Conductance)
Resistance thermometers, 6, 15, 16 (see also Thermometers)
Rhodium, 7, 40–42, 154, 155 (see also Wire, thermocouple)
Rosin, 129, 138
Rotary table, 118, 119
Rubber, 86–89, 98, 129, 134, 138, 157

Runout, in drilling, 108, 114, 123, 124, 126 (see also Drills)
Ruthenium, 42 (see also Wire, thermocouple)

Safety factor, 71
Scale, temperature (see Temperature scale)
Schlieren photography, 20
Screw-plug junctions, 90, 91, 96–98 (see also Junctions)
Scriber, 119
Sealing wax, red, 138
Sealstix, 138
Sensitive element, 14, 31, 68–83
Sensitivity, circuit arrangements for, 61, 65
Series, thermocouples in, 59, 65
Sheathing, 41, 74, 81, 86, 89, 95, 130, 131, 135, 143, 145, 151, 153, 155, 157, 158, 162, 166
Shellac, 138
Signal, 64, 65
Silica, 136
Silicon, 132
Silicon carbide, 42, 124, 125, 135, 137
Silicone, 96, 130, 134, 135, 139, 156, 157 (see also Coatings)
Silk, 130
Sillimanite, 101
Silver, 6, 42
Silver soldering, 94–96, 111, 124, 125, 133, 137
Soapstone, 41, 109, 132, 136
Soldering, 47–49, 74, 75, 92–98, 124, 125, 133, 137, 157 (see also Brazing)
   contamination by, 47, 48 (see also Contamination)
   fluxes for, 39, 48, 49, 124, 125, 133
"Spaghetti" (see Sheathing)
Spark-drilling, broken drill removal by, 126
Spurious emfs, 37–39, 41, 48, 55–58, 61, 65, 66, 81, 90, 145, 146 (see also Parasitic emfs)
Stainless steel, 89, 96, 126, 132
Standards, thermocouple-calibration, 45
Steam point, 4, 6, 11 (see also Temperature scale, fixed points for)

SUBJECT INDEX

Steatite, 136
Sulfur, 6, 132 (*see also* Temperature scale, fixed points for)
Superconducting thermometer, 16 (*see also* Thermometers)
Switches, 37, 58, 66, 71 (*see also* Circuits, thermocouple)
Systematic errors, 25, 27, 28, 81–83 (*see also* Errors)

Talc, 136
Tantalum, 124, 132
Teflon, 134
Telemetering, 64
Temperature, gradient of, 76–83, 142, 159–167
  inconstant, 69
  measurement of, at a point, 3, 68–83, 100, 101, 108, 109, 145, 146, 148, 159–161
  basic complexities, 3
Temperature scale, absolute, 2, 4
  Celsius, 4
  concept of, 1
  Fahrenheit, 1, 5
  fixed points for, 1, 5–7, 10, 11, 45, 59, 60
  gas, 2
  International, 4–10, 46
  "International" Vapor Pressure, 8, 9
  Kelvin, 4
  magnetic, 9
  Provisional, 8, 47
  Rankine, 5
  thermodynamic, 2, 4
    below 11°K, 8, 9
  thermodynamic centigrade, 4, 5
Tests, insulation, electrical, temperature resistance of, 145–147
  thermocouple installations, accuracy of, 145, 146
Thermistor, 16 (*see also* Thermometers)
Thermocouple thermometer (*see also* Circuits; Junctions; Wire)
  calibration of, 6, 7, 44–46, 100, 101 (*see also* Wire)
  characteristics of, 16, 17, 20, 84

Thermocouple thermometer, circuits for (*see* Circuits, thermocouple)
  classification of, 14, 16, 17
  installation of, 68–106, 142–166 (*see also* Junctions)
    small-scale, 153–157
    spark-plug-electrode, 155, 156
    tests for accuracy of, 145, 146
  junctions for (*see* Junctions)
  laws of, 37
  leads for, 31, 36, 71–74, 128, 154–158, 166 (*see also* Circuits; Junctions)
    anchoring of, 144, 145
    base-metal, 35, 36
    emerging separately, 151, 152, 163–166
    emerging together, 142–151, 161–163
  parallel connection of, 59 (*see also* Circuits
  power of, definition, 34, 35
  tables, 34, 35, 38–43, 81–83
  series connection of, 59, 65 (*see also* Circuits)
  wire combinations for (*see* Wire, thermocouple)
Thermoelectric power, 34, 35, 38–43, 81–83 (*see also* Thermocouple thermometer)
Thermoelectricity, laws of, 37 (*see also* Thermocouple thermometer)
Thermometers (*see also* Pyrometer)
  alcohol-in-glass, 1, 14, 15
  bimetallic, 16
  Bourdon, 14, 15
  classes of, 14, 15
  definition of, 14
  gas, 1, 5, 8, 9, 14
  liquid-filled, 1, 10, 11, 14, 15
  mercury-in-glass, 1, 10, 11, 14, 15
    calibration of, 10, 11
    effect of external pressure on bulb, 10
    emergent-stem correction for, 11
    hysteresis effect in, 10
    progressive changes in, 10
  radiation, 7, 13, 14, 17, 18, 46 (*see also* Pyrometer)
  resistance, 6, 15, 16

SUBJECT INDEX

Thermometers, resistance, electrolytic, 16
  platinum, 6
  superconducting, 16
  thermistor, 16
  thermocouple (*see* Thermocouple thermometer)
  vapor-pressure, 15
Thermopile, 59, 65 (*see also* Thermoelectric thermometer)
Thermosetting cements, 138, 139 (*see also* Cement)
Titanium, 124
Thoria, 137
Triple point, 7 (*see also* Temperature scale, fixed points for)
Tube-wall thermocouple, boiler, 84, 85, 92–96 (*see also* Junctions, tube-wall)
Tungsten, 42, 46, 124, 125, 133
Turpentine, 116
Twist drills, 109–124 (*see also* Drills)

Vanadium, 124
Varnish, spar, 129, 130 (*see also* Coatings)
Vaseline, 129
Velocity of sound, method of, 13, 20
Vernier depth gage, 121
Vernier height gage, 119
Vibrations, isolation of instrumentation from, 66
  ruggedness with respect to, 31, 66, 67, 70, 94, 128, 137, 143, 147, 150

Water-repellent coatings, 39, 130, 131, 134, 143, 157, 158, 162 (*see also* Coatings)

Welding, 49–55, 74, 75, 89–91, 133, 135, 137, 155
Wire, thermocouple, bismuth against bismuth alloy "B," 42
  calibration of, 6, 7, 44–46, 100, 101
  carbon against silicon carbide, 42
  chromel P against alumel, 40, 41, 86
  chromel P against constantan, 42
  constantan against silver, 42
  copper against constantan, 39–41, 43, 86, 88, 93, 94
  gold-cobalt against silver-gold alloy, 43
  homogeneity of, 36–38, 43–45, 48, 49, 90, 131, 133
  iridium against rhodium-iridium alloy, 42
  iridium against ruthenium-iridium alloy, 42
  iron against constantan, 40, 42, 81–83, 94, 143–147, 150, 151, 162–164, 166
  platinum against platinum-rhodium alloy, 7, 40, 41, 154, 155
  silver-gold against copper-iron alloy, 43
  testing of, 43, 44
  tube-enclosed, 90, 91
  tungsten against iridium, 42
  tungsten against tungsten-molybdenum alloy, 42
Wood, 86, 109

X-rays, 19

Zirconia, 46, 137
Zirconium silicate, 137